Planetary Systems

Volume 1 – Classic Systems

Klaus Piontzik, Claude Bärtels

© Klaus Piontzik, Claude Bärtels 2014
Planetary Systems - Volume 1

Production and publishing:
BoD – Books on Demand GmbH, Norderstedt

ISBN 978-3-7357-3854-7

Planetary Systems

SUMMARY

	Page
Part 1 – Basics	10
Introduction	10
1 – Theoretical approach	12
1.1 Classic physical systems	14
1.1.1 Geologic layers	14
1.1.2 Atmospheric layers	14
1.1.3 Earth magnetic field	15
1.1.4 Electric field of the earth	15
1.2 Polyhedron models of the earth	16
1.2.1 Polyhedrons and grids	17
1.3 Geometrical structuralisation	19
1.4 Assertions for an oscillation structure	19
2 – Approach for an oscillation model	20
2.1 Spherical harmonics	22
2.2 Addition and multiplication of waves	24
2.2.1 Zero-grid	24
2.2.2 Pole forming	24
2.2.3 Grid forming	25
2.3 Huygens-Fresnel principle	28
2.4 Basic oscillations	30
2.5 Radial structure	32
2.5.1 Stratification	33
2.5.2 Layer Calculation	34
2.5.3 Normalisation	38
2.6 Radial stationary waves	40
2.7 Stratification structure	42
2.8 Spatial grid	43
2.9 Spatial oscillation structure	44
2.10 Global net grids	45
2.11 General attempt	46
2.11.1 Angle part	47
2.11.2 Radial part	48
2.11.3 General remarks	49

Page

Part 2 – Applications 1   50

3 – Frequencies of the earth   51
3.1   Sferics   52
3.2   Basic hull   55
3.3   Table of the earth layers   56
3.4   Analysis procedure   57
3.5   Geologic layers   58
3.6   Geologic layers and Laplace   61
3.7   Layers of the atmosphere   68
3.8   Layers of the atmosphere and Laplace   73
3.9   Planetary oscillation systems   80
3.10 Layers and frequencies   82
3.11 Schumann frequency   84
3.12 Summary   85

Part 3 – Applications 2   86

4 – Earth magnetic field   87
4.1   Gauß and Weber   89
4.2   Measuring stations   90
4.3   Total intensity – WMM 2005   91
4.4   Temporary stability   93
4.5   Fourier analysis of the earth magnetic field   95
4.5.1 Fourier analysis   96
4.5.2 Quantitative Fourier analysis   99
4.6   Further evaluations   100
4.6.1 Three-axle ellipsoid   101
4.6.2 Grid ZS   102
4.6.3 Tesseral field   103
4.6.4 Huygens source points   104
4.6.5 Summary   105
4.7   Huygens source points of the earth field   106
4.7.1 Ideal source points structure   106
4.7.2 Real source points structure   108

		Page

5 – Generating and generated elements — 109
- 5.1 Core balls — 110
- 5.1.1 Case 1 — 110
- 5.1.2 Case 2 — 111
- 5.2 Creation of geological layers — 113
- 5.3 The electric field of the Earth — 116
- 5.4 One oscillation structure — 118
- 5.5 Sub structure — 119

Part 4 – Demonstrability — 120

6 – Measuring of magnetic waves — 121
- 6.1 Classic Hall sensor — 121
- 6.2 New functionality — 123
- 6.3 Circuit to the measuring procedure — 126
- 6.4 Experimentum crucis — 127
- 6.5 Synthesis — 128

7 – Converting a numerical sequence in an e-function — 130
- 7.1 Numbering — 130
- 7.2 Logarithmic — 131
- 7.3 Linearization — 133
- 7.4 Determination of the approximation line — 133
- 7.5 Determination of the e-function — 135
- 7.6 Determination of a new numbering — 136
- 7.7 Global Scaling — 137

8 – Concentric arrangements — 140
- 8.1 The Sun — 140
- 8.2 The orbits of the planets — 147
- 8.3.1 Moons of the planets — 154
- 8.3.2 Moons of Mars — 159
- 8.4 Rings of planets — 161
- 8.4.1 Rings of Saturn — 161
- 8.4.2 Rings of Jupiter — 166
- 8.4.3 Rings of Neptune — 167
- 8.4.4 Rings of Uranus — 168

		Page
8.4.5	Rings of Rhea	170
8.5	Satellite galaxies of the milky way	171
8.6	Planetary nebulae	175
8.7	Layer of the earth	177
8.8	Fruits and flowers	183
8.8.1	Peach	183
8.8.2	Narcissus	185
8.9	Result	188

9 – Epilog   190

Bibliography   194

Picture credits   198

Part 1 – Basics

Introduction

This book is the essence and progression of two presentations, the author K. Piontzik has held on 3/14/2009 to the association for the support of the geobiology in Brügge and on the spring conference of the research group for geobiology in 4/24/2009 in Eberbach.

The present material represents a progression and completion of the book "Gitterstrukturen des Erdmagnetfeldes". The basic material (about 60%) of the book "Gitterstrukturen des Erdmagnetfeldes" is also on the Internet in English accessible at: www.pimath.eu.
Seen from today's perspective „lattice structures of the earth magnetic field" delivers a collection of facts and basic informations. A closed homogeneous model is recognizable in attempts, but the golden threat is still missing.
Now in cooperation with the biophysicist Dr. Claude Bärtels the new work „Planetary systems" forms a closed uniform working hypothesis, with that the physical layer structures of the earth (geologic bowls, atmospheric layers, earth magnetic field, electric earth field) can be explained.

This book is a working hypothesis, which can be falsified after the today's epistemology by Popper.
From it the information of a physical measuring method occurs which represent the **experimentum crucis** for this working hypothesis.

Overall, the model shown here, represents a holistic approach on an oscillation base that explains several structures of the earth.

What is valid on the earth, must be valid also "in" the earth and with it is valid both on the large and small scale. What means for the authors that the macroscopic oscillation structures have her correspondences also in the (sub) microscopic (atomic) area.

The course of the development in the last both years has shown, that the whole themes about stratifications, oscillations and grids can be attributed to a central concept with which all phenomena can be explained: it is the concept of the **planetary system**.

1.0 – Theoretical approach

The question which raises here is: What is to be understood by a planetary system? In addition the concept must be examined such closer.

Planetary means a global earth phenomenon. And the concept system implies that a certain order exists.
Planetary systems are so global or earth-enforcement structures. Two systems are recognised in the following picture: the earth's surface even with her lying underneath geologic bowls and the atmosphere with her stratifications.

Illustration 1.0 – the earth

Definition 1.0.1: **Earth System**
= All energy and matter that is present in the area of planet Earth and its neighbourhood.

The neighbourhood of the planet Earth can be explained by a spherical space with a radius of 2 or 3 Earth radii. Here, the concept of neighbourhood is reminiscent of the definition from the topology.

Definition 1.0.2: **Planetary System**
= a global elementary subsystem of the earth system with a geometrical structuralisation.

The geometrical structure would have to be defined even closer. This happens in the following chapters.

Definition 1.0.3:

Structures that are considered as planetary systems:

1) Geologic bowls
2) Atmospheric layers
3) Earth magnetic field
4) Electric field of the earth

5) Polyhedron models of the earth

In the following we will have a look at the listed systems.

1.1 – Physical systems

1.1.1 – Geologic layers

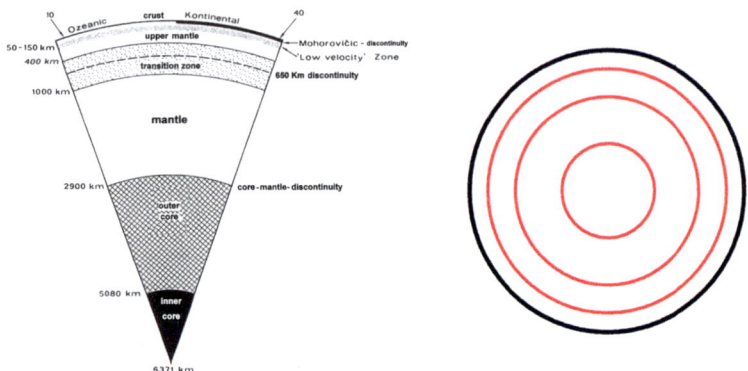

Illustration 1.1.1 – geologic layers

The geometrical structuralisation of the geologic bowls exists as **concentric balls** in the earth.

1.1.2 – Atmospheric layers

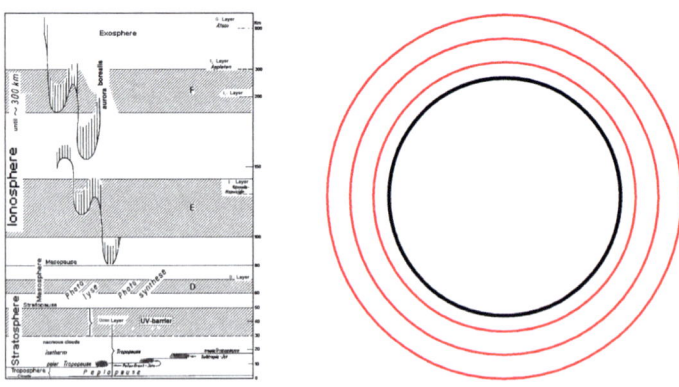

Illustration 1.1.2 – atmospheric layers

The geometrical structuralisation of the atmospheric layers exists as **concentric balls** round the earth.

1.1.3 – Magnetic field of the earth

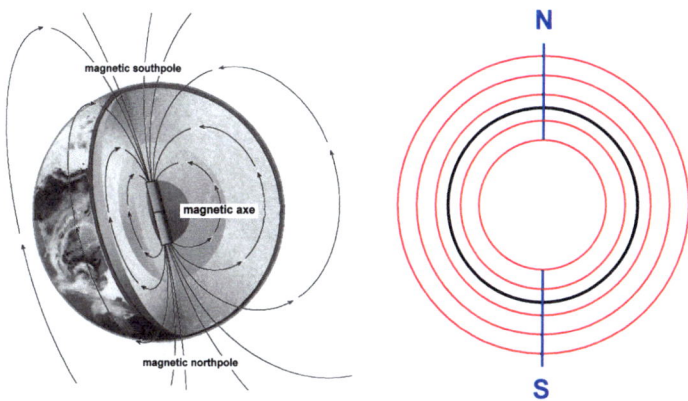

Illustration 1.1.3 – magnetic field of the earth

The geometrical structuralisation exists (simplisticly) as **concentric balls** in the earth and round the earth. **Polar** (north south pole) and **radial** structures (magnetic flux density) still appear.

1.1.4 – Electric field of the earth

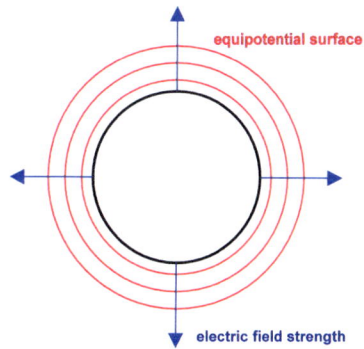

Illustration 1.1.4 – electric field of the earth

The geometrical structure exists as **concentric spheres** round the earth. **Polar** (plus minus poles) and **radial** structures (electric field strength) still appear.

1.1.5 – Result

Geologic spheres, atmospheric layers, earth magnetic field and the electric field own a geometrical structuralisation which exists of **concentric spheres**.
The geologic bowls lie within the earth. The atmospheric layers and the electric field lie outside, around the earth.
The earth magnetic field exists in the earth as well as round the earth. With magnetic field and electric field **radial** and **polar** structures still appear.

1.2 – Polyhedron models of the earth

At the end of the 19th century the geologists **W.L. Green** and **A. de Lapparent** compared the shape of the earth with a tetrahedron.
A similar comparison did in the sixties of the twentieth century **B.L. Litschkow** and **N.N. Schafranowski** with an octahedron. Litschkow published a little later the model of a dodecahedron and an icosahedron for the earth shape.
In 1974 **Nikolai F. Gontscharow, Wjatscheslaw S. Morosow** and **Walerij A Makarow** published in the Russian magazine "Chimi-ja i Zisn" (chemistry and life, No. 3, March) the model of a dodecahedron of the earth.

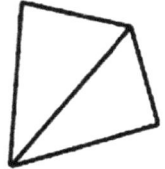

Tetrahedron
end of 19. century
W.L. Green
A. de Lapparent

Octahedron
in the 60. years of 20. century
B.L. Litschkow
N.N. Schafranowski

Illustration 1.2.1 – polyhedron models of the earth

Icosahedron
In the 60. years of 20. century
Litschkow

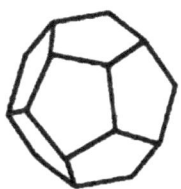

Dodecahedron
1974
Nikolai F. Gontscharow
Wjatscheslaw S. Morosow
Walerij A. Makarow
1999 – S. Prumbach

Illustration 1.2.2 – polyhedron models of the earth

These polyhedron models of the earth were developed primarily by geologists. There is only one body missing to receive a certain set from bodies.

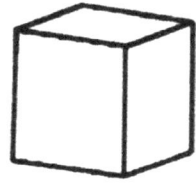

There is only the cube missing to complete the **Platonic solids**.
Platonic solids are regular bodies which are built up from regular bases. There exist only five Platonic solids as shown here.

Illustration 1.2.3 – cube

1.2.1 – Polyhedrons and grids

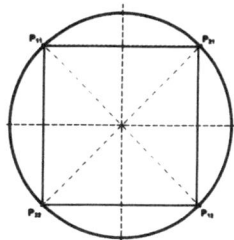

The corner points of the polyhedrons lie on the **infolding** sphere. All edges are transferred to the ball surface. In contrast to all other planetary systems lines with **finite** lengths occur here, due to the structure of the polyhedron.

Illustration 1.2.4 – polyhedron in cross section

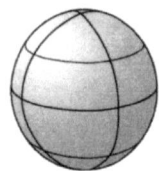

You can **project** the edges or corners of the polyhedron on the enveloping ball.
The best way is to depict all corners of a polyhedron as intersections of circle of latitude and meridians, such like the geographical system.

Illustration 1.2.5 – Gitterbildung

The points of intersection of the resulting grid correspond to the corners of the polyhedron. And the edges of the polyhedron are usually on the circles of latitude or meridians.

The octahedron is the only platonic solid in which all edges of the polyhedron and grid lines fully match. The grid is used by three, each perpendicular on each other standing circles formed. As a result, the spherical surface is decomposed into eight equal parts.

The cube has a decomposition which is formed by four circles and also forms a symmetrical grid. All horizontal edges of the cube are identical with the circles of latitude. All vertical edges of the cube are included with the meridians as parts.

The tetrahedron has the same separation as the cube, a tetrahedron can be represented as an inside body of a cube. Edges of the polyhedron occur but they are not directly represented in the coordinate system.

Another possibility for the tetrahedron is to see this as a triangular pyramid. So one would get three meridians, and a parallel, but also an irregular grid.

Also the icosahedron provides a regular grid with five meridians and two circles of latitude. Edges of the polyhedron occur, but that are not directly represented in the coordinate system.

Only the dodecahedron builds irregular distances with five meridians and four circles of latitude.

All together the following statement can be formulated:

1.2.2 - Theorem: Polyhedron ⇔ Grid

Polyhedrons are equivalent to grids on a ball surface.

1.3 – Geometrical structuralisation

For the systems to be looked exist, according to 1.1 and 1.2, following geometrical structuralisations:

1.3.1 - Concentric balls	ball shape, concentricity
1.3.2 - Polar structures	north-south poles plus-minus poles
1.3.3 - Radial structures	magnetic flux density electric field strength
1.3.4 - Grids	Polyhedrons

1.4 – Assertions for an oscillation structure

1.4.1 - Definition: Oscillation structure
= A system of patterns of vibration or oscillating structures that are related to each other.

This vibration structure form usually spatial structures, whose basic patterns include geometric solids, both have harmonical ratios to each other.

According to definition 1.0.2 is valid:

Planetary system = a global elementary subsystem of the earth system with a geometrical structuralisation.

1.4.2 - Assertions:

1.4.2.1 - A planetary system can be explained by a spatial **Oscillation structure**.

1.4.2.2 - All planetary systems can be explained by **just one** spatial oscillation structure.

2.0 – Approach for an oscillation model

The aim of this chapter is the description of basic mathematical and physical terms and conditions, that serve the development of an equation for an oscillation structure and allow a quantification of the model. The approach is based on **oscillations around a ball**.

Examples for oscillation possibilities:

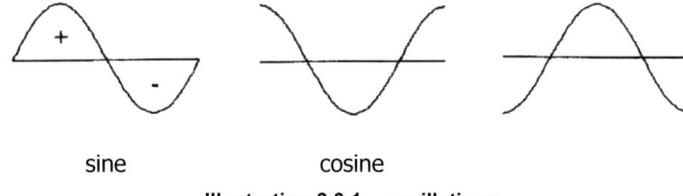

      sine   cosine

Illustration 2.0.1 – oscillations

sine or cosine = oscillation = wave

Applies to physical oscillations:

2.01 - Equation: $f \cdot \lambda = c$

(Frequency multiplied with wavelength is equal to speed of light)

How to get vibrations around a ball ? - Analogous to the Bohr model of the atom, if it contains the surrounding Electron as a wave by de Broglie:

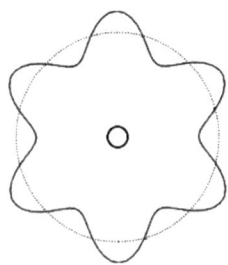

Illustration 2.0.2 – oscillations around a ball

It fits only an integer number of oscillations around the globe.

2.02 - Equation: $\quad\quad\quad n\cdot\lambda \Leftrightarrow 360° = 2\pi \quad\quad n\in N$

The wavelength is proportional to the circle angle alpha:

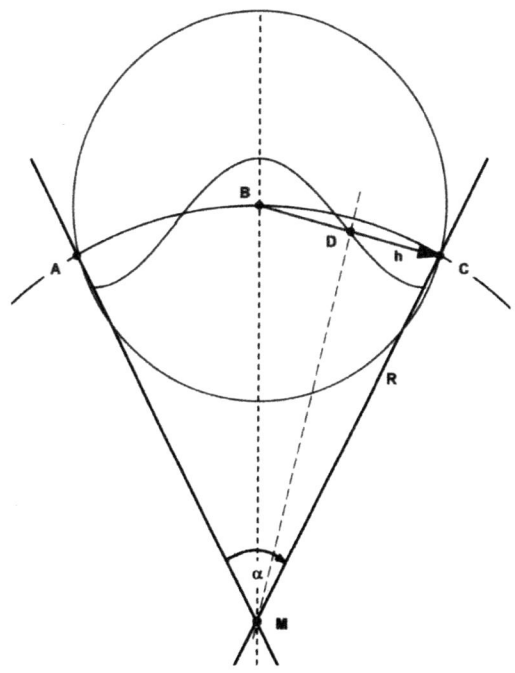

Illustration 2.0.3 – wave length and circle angle

2.03 - Equation: $\quad\quad\quad \lambda \Leftrightarrow \alpha$

Condition for **n** vibrations around a globe:

2.04 - Equation: $\quad\quad\quad n\cdot\alpha = 2\pi \quad\quad n\in N$

Theoretically, the following form is possible:

2.05 - Equation: $\quad\quad\quad n\cdot\alpha = 2\pi\cdot m \quad\quad m,n\in N$

Here, the oscillation circle does not close after one revolution, but only take **m** turns.

2.1 – Spherical harmonics

A standing wave around a sphere can be interpreted as a stationary state. Thus, each state of a wave is spatially fixed. The question now is: how many waves fit around a globe?
In classical mechanics, degrees of freedom is the number of freely selectable, independent movements of a system
A rigid body in space has a degree of freedom f = 6, because you can move the body in three independent directions and rotate in three independent planes.
Because a sphere is rotationally symmetric, so rotations are irrelevant. A ball has therefore 3 freedoms regarding a wave propagation. Therefore, three independent waves around the globe are possible.
Due to the spherical shape, the three freedoms can be represented as spherical coordinates.

Example earth:

1) A wave starts from North Pole via the South Pole to the North Pole
2) The second wave runs around the equator
3) The third standing wave runs radially - based from the center

There exists a mathematical concept namely for the first two examples, that is suitable for a representation - the **spherical harmonics**.
Standing waves on the surface of a sphere can be treated as spherical harmonics. There are 3 types:

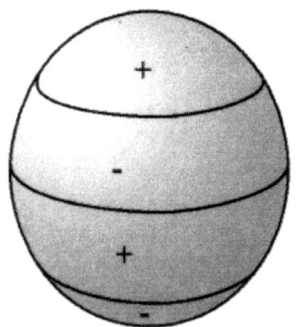

Zonal spherical harmonics only depend on the latitude. $\sin\varphi$
$\cos\varphi$

Illustration 2.1.1 – zonal spherical harmonic

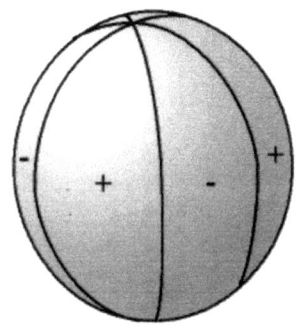

Sectorial spherical harmonics depend only by the degree of longitude.

$\sin\lambda$
$\cos\lambda$

Illustration 2.1.2 – sectorial spherical harmonic

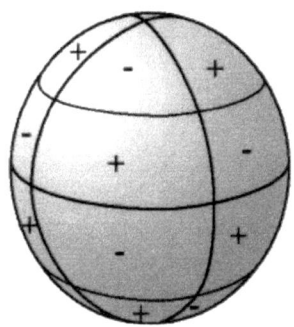

Tesseral spherical harmonics depend of the latitude and the longitude.

$\sin\varphi \cdot \sin\lambda$
$\sin\varphi \cdot \cos\lambda$
$\cos\varphi \cdot \sin\lambda$
$\cos\varphi \cdot \cos\lambda$

Illustration 2.1.3 – tesseral spherical harmonic

2.1.1 - Definition: tesseral spherical harmonic
= product of two oscillations
= 2 perpendicular standing waves
= **grid**

Comment:

A complete route in terms of square grid on a sphere can not materialize. Only grid systems are developed, which are designed as the geographic grid system. There are always **two** poles. The corresponding "meridians" and "circle of latitude" then make the **grid**.

2.2 – Addition and multiplication of waves

spherical harmonics can present itself as two sine- or cosine waves, which are perpendicular to one another and additative or multiplicative overlap.

2.2.1 - Zero-grid

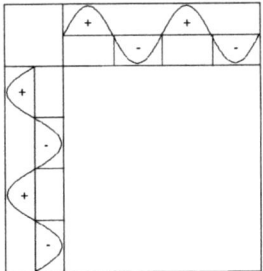 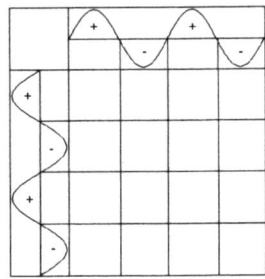

Illustration 2.2.1 – zero-grid

The zero points of both waves are transferred on the consideration level, as shown in Illustration 2.2.1 right.
This is a tesseral spherical harmonic with: $G_0 = \sin\alpha \cdot \sin\beta$

Two vertical waves can be added by following **qualitative** rules:

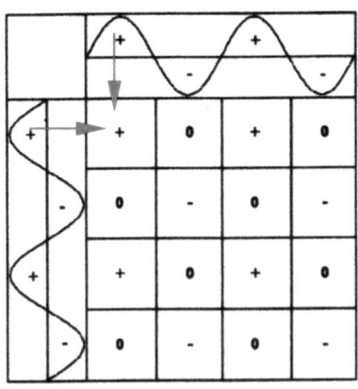

Illustration 2.2.2 – multiplication

2.2.2 – Pole forming:

1) + and + is +
2) – and – is –
3) + and – is 0

This procedure results in fields with different portents or different states. There exist three oscillations states: **positiv(+), negativ(-), neutral(0)**

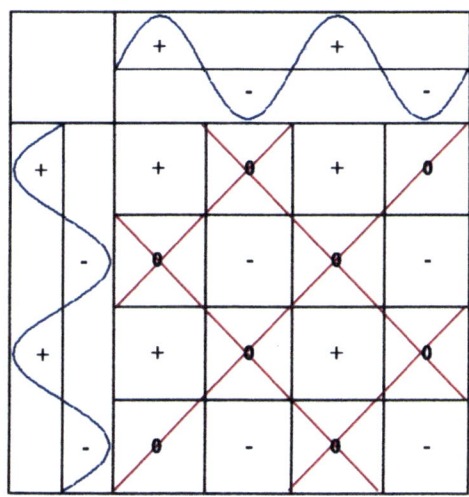

Illustration 2.2.3 – generated grid

2.2.3 – Grid forming:

It is striking that all zero fields are **diagonal** to each other.

Connecting the zero fields as shown in Illustration 2.2.3.

2.2.4 - Definition: The (red) grid-like structure is then called **basic field** or **grid** or **generated field**.
The (blue) producing waves are called **basic oscillations**.

The basic field:
$$G = \sin\alpha + \sin\beta$$

While the mathematical concept of spherical harmonics does not ask for the cause of the oscillation field, so the underlying waves must be included on the physical examination.

The term of the basic field does this. **The basic field is defined by the basic oscillations.**

The name **basic field** is regarded as a physical equivalent to the mathematical concept of tesseral spherical harmonics.

The multiplication of the waves works as seen first in a discreet manner and way.
A continuous point multiplication of two perpendicular to one another standing waves result in grid pattern, with alternating polarities of the grid fields, as in the illustration 2.2.4 shown.

This shows that the field maxima and minima field points occur in the middle of the squares, while the lines consist of **zero values**.
The field maxima are there as hills, while the valleys are formed by the field minima.

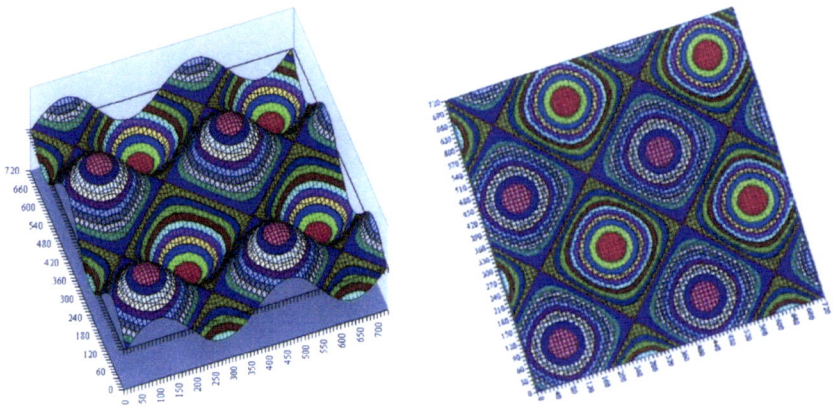

Illustration 2.2.4 – generated grid

On the bottom left and right in the left image the generating oscillations are visible. It is also visible that two generated grid fields result in a (generated) oscillation.

2.2.5 - Definition: **Grid**
= two-dimensional oscillation structure

It is possible here to draw a second grid. It is the **maximum grid**. It combines the field maxima and minima, so the top of the hill and the valley bottoms in the illustration 2.2.4 and represents the extreme gradient of the field.

This allows **two** perspectives of the grid:

2.2.6 - Definition:

1) the generated grid is described in the plane of **basic oscillations** with: **G = sinα + sinβ**

2) The generated gird is described in the grid plane itself.
   G = sinφ · sinλ

Both coordinate systems differ, they are turned against each other by 45 degrees.

For Illustration:

A physical analogue to the ground field are the

Chladni-figures

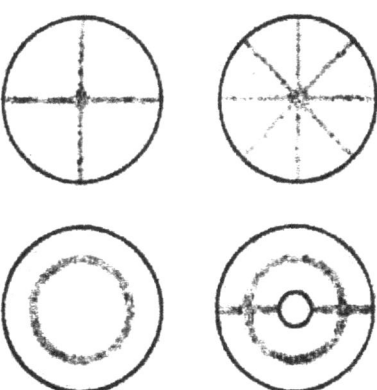

Illustration 2.2.5 – Chladni-figures

For producing the figures, a metal plate is sprinkled with sand and then moved to vibration.
When resonance occurs at certain frequencies with the natural frequencies of the plate, than the entire plate begins to oscillate.
Thus, the sand exactly in the places stays where the amplitude of the vibration is zero.

2.3 – Huygens-Fresnel principle

According to chapter 2.1 a ball has three degrees of freedom f = 3. Two degrees of freedom could be covered through the use of spherical harmonics.
The third freedom, the **radial** direction, is still missing. This requires the understanding of a physical representation with which the expansion of physical waves can be described: **the Huygens-Fresnel principle**.

The huygensche principle emanates from a source **S**, which generate wave fronts uniformly in all directions. To get the resulting wave front in the point **P**, it is not necessary to consider the entire spread of **S**.
The **Huygens-Fresnel principle** says that every point (**O**) from a wavefront can be considered as a starting point of a new wave, the so-called **elementary wave**.
The location of the resulting wavefront in **P** arises from superposition of all elementary waves.

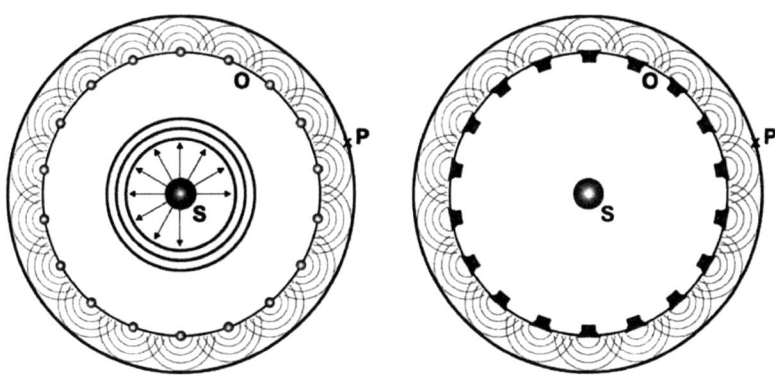

Illustration 2.3.1 – Huygens-Fresnel principle

The **origins of the waves** (O) deliver the resulting wave front (P) by superposition of **elementary waves**.

In **two** dimensions, elementary waves are **circular**.
In **three** dimensions, elementary waves are **spherical** in shape.

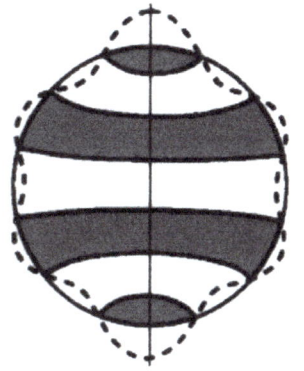

Application of the Huygens-Fresnel principle for a standing wave around a globe:

**Extreme values of the wave
= sources
= waves origins**

1 wave = 2 sources

Illustration 2.3.2 – wave origins

The following image shows the oscillation state for a wave by the Huygens-Fresnel principle with a maximum (wave peak) as a source.

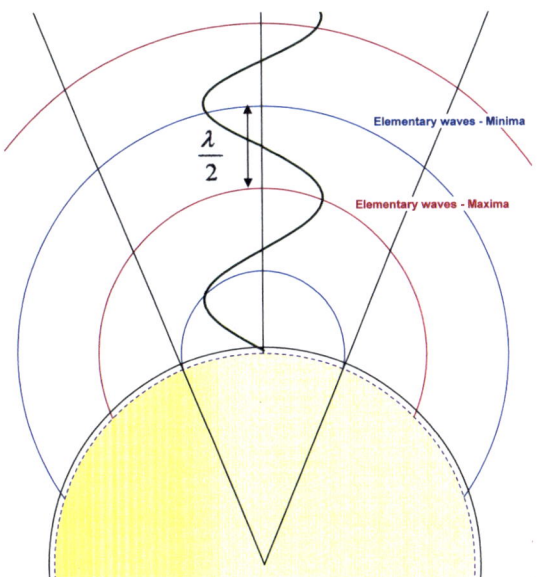

Illustration 2.3.3 – elementary waves

Since a wave consists of maxima and minima, the elementary waves also form minimal fronts (blue) and maximum fronts (red). Between the extreme fronts course zero fronts exist.

2.4 – Basic oscillations

Now the previous considerations allow a first calculation of a wave that builds up, as the earth, around a sphere. Still, it must be considered that the wave propagation takes place **linear** and **not** along the curved surface of the ball.

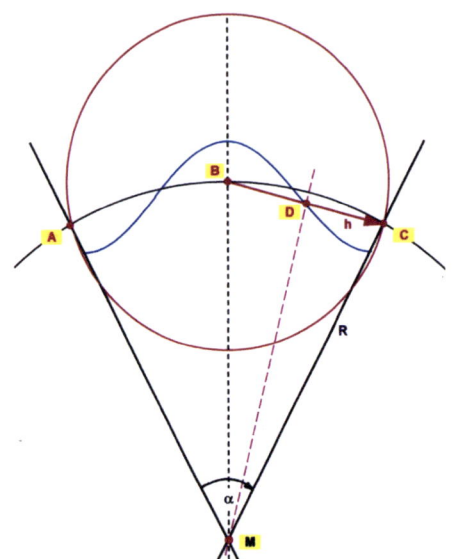

After the **Huygens-Fresnel principle** the points **A**, **B**, **C** serve so as **source points** of the standing wave (blue)

Electromagnetic waves propagate **spherically** from one point. (red)

If point **B** is the starting point so the distance **BDC = h** is the way of the wave.

Illustration 2.4.1 – wave propagation

Because there is a **stationary state**, it is sufficient to look at the way of the wave **from one source to the next source**.

This makes: $$h = \frac{\lambda'}{2}$$

The triangle of **MCD** in the picture 2.4.1 is rectangular in the point **D** and apply:

$$\sin\frac{\alpha}{4} = \frac{\frac{h}{2}}{R} = \frac{h}{2R}$$

Shift the equation to **h**:

$$h = 2R \cdot \sin\frac{\alpha}{4} = 2R \cdot \sin\frac{\pi}{2n} = \frac{\lambda'}{2}$$

Since now all sizes are known, the wavelength can be determined:

2.4.1 - Equation: $\quad \lambda' = 4R \cdot \sin\frac{\pi}{2n}$

In the equation is **R** the radius of the Earth, **n** is the number of waves.

This results in the **frequency equation for basic oscillations**:

2.4.2 - Equation: $\quad f = \dfrac{c}{4R \cdot \sin\dfrac{\pi}{2n}} \quad$ n = 1,2,3,4,...

The **basic frequency** is for **n = 1**:

2.4.3 - Equation:

$$f_o = \frac{c}{4R}$$

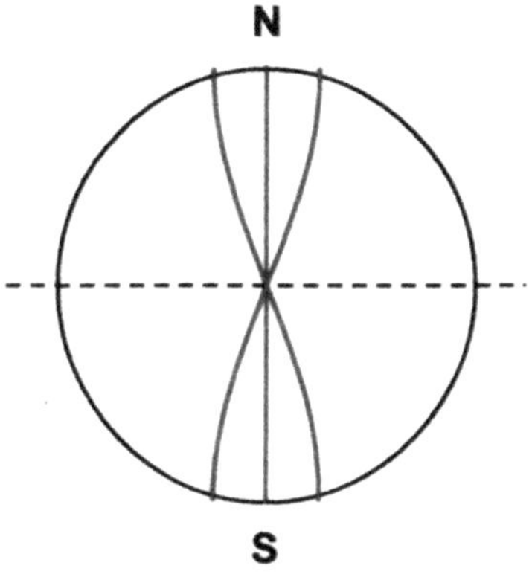

Illustration 2.4.1 – basic frequency

This is also the oscillation of a bar with free ends, which length is equal to the diameter of the ball.

2.5 – Radial structure

The picture shows the oscillation state of **n** waves, when the Huygens-Fresnel principle is applied to all sources. The maximum fronts are red, the minimal fronts are blue.

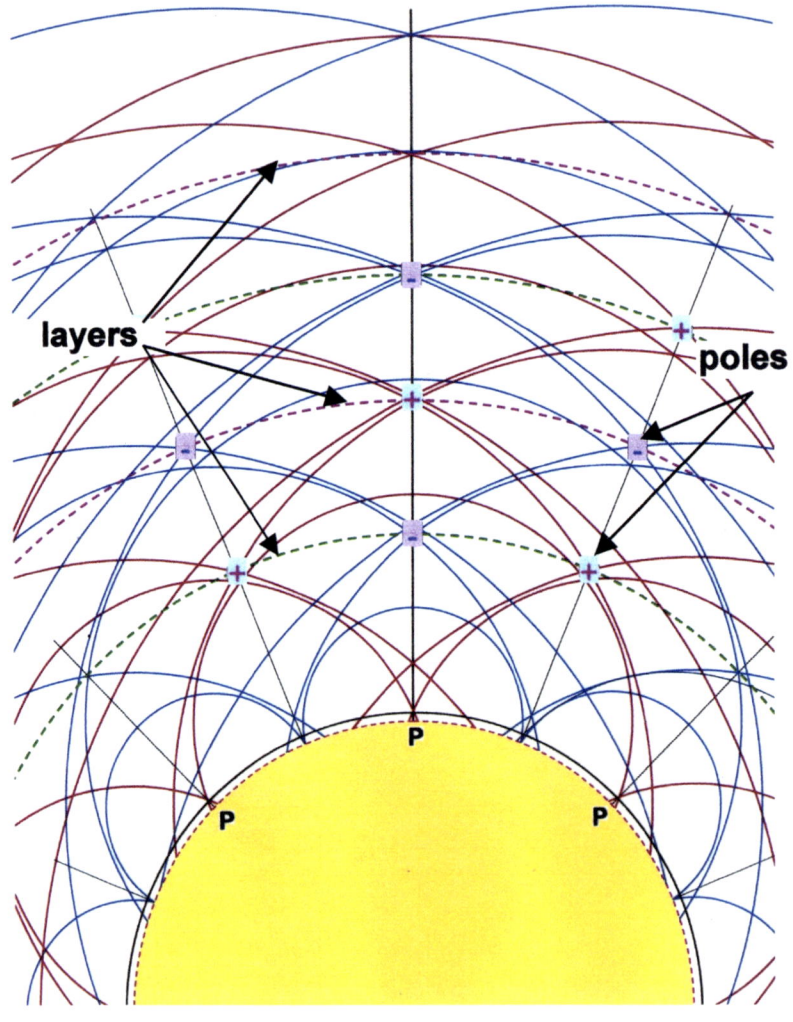

Illustration 2.5.1 – superposition of waves

2.5.1 – Stratification

The superposition of elementary waves follows the same rules as described in section 2.2.

2.5.1.1 - Definition: Pole forming

Plus poles are formed by the superposition of the positive elementary waves.

Minus poles are formed by the superposition of the negative elementary waves.

Zero poles are formed by the superposition of positive and negative elementary waves.

2.5.1.2 - Definition: Layer formation

By the superposition of elementary waves, **poles** are formed (+,-, 0) on **concentric spherical layers** in the following **Layers = L** called.

2.5.1.3 - Definition: Wall forming

Zero poles arise between these layers, which also lie on **concentric spherical layers** and are called in the following **zero walls**.

By the **interference** of different waves, raises oscillation maxima and minima or **oscillation layers** (See illustration 2.5.1) that envelope the earth **spherically**.

The layers on which the poles are can be calculated. The following derivation use a geometric approach to **calculate the distance of a layer** to the earth center.

2.5.2 – Calculating the layers

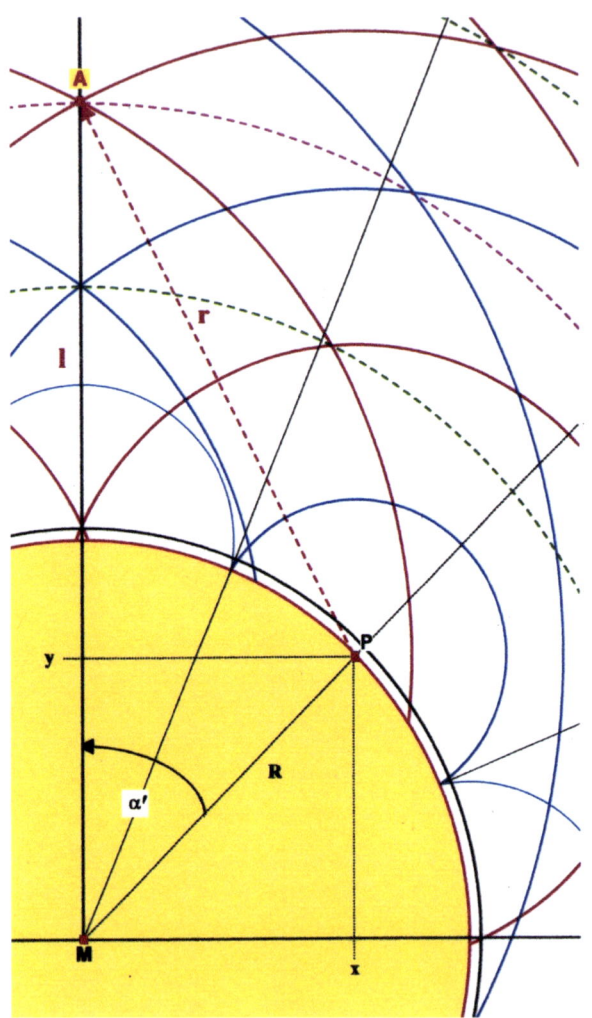

Illustration 2.5.2 – distance calculating of the magnetic layers

To determine is the center distance **MA=l'** of a magnetic layer.

One has for the line **MA**:
$$l' = y + l$$
and
$$l = \sqrt{r^2 - x^2}$$

There are the relations: $\quad x = R \cdot \sin\alpha' \quad$ und $\quad y = R \cdot \cos\alpha'$

With the basic hull radius **L₀=R=6355758,426 m**

Based on Chapter 2 each oscillation can be allocated to a certain angle α:

The base angle is: $\quad\quad\quad \alpha = \dfrac{2\pi}{n} \quad$ mit $\quad n \in N_0$

n is the number of oscillations around the circle. Alternatively, here also the application of equation 2.05 would be possible.

The distance from a starting point of the field to the next is half of the angle alpha. Taking into account all source points, so they occur in integer multiples of half basic angle.
It is so true:

Multiple of the angle: $\quad \alpha' = m \cdot \dfrac{\alpha}{2} \quad$ mit $\quad m \in N_0 \quad$ und $\quad m \leq n$

In the illustration 2.5.2 one has for the source point **P**: m=2

The route of **PA = r** corresponds to the way the wave travels. From the source point **P** starting up to the oscillation layer (point A), which is to determine.
Caused by the stationary state, maximum fronts and minimum fronts arises, with regular intervals spherically around the source point.
After chapter 2 corresponds a distance to half a wavelength. Hence, the way of the wave up to the oscillation layer is always an integer multiple of half a wavelength:

The distance travelled by the wave: $\quad r = k \cdot \dfrac{\lambda}{2} \quad$ mit $\quad k \in N_0$

In illustration 2.5.2 applies to the wave way **PA**: k=4

Applies to the wavelength: $\lambda = 4R \cdot \sin \dfrac{\pi}{2n} = 4R \cdot \sin \dfrac{\alpha}{4}$

Inserting all treated terms in the equation for **l'** leads to the following conclusions:

2.5.2.1 - Equation:

$$l' = R \cdot \left(\cos\alpha' + \sqrt{4k^2 \cdot \sin^2 \dfrac{\alpha}{4} - \sin^2 \alpha'} \right)$$

Are alpha and Alpha line replaced by the corresponding terms from the previous considerations, so is the following end equation:

2.5.2.2 - Equation:

$$L = R \cdot \left(\cos \dfrac{m\pi}{n} + \sqrt{4k^2 \cdot \sin^2 \dfrac{\pi}{2n} - \sin^2 \dfrac{m\pi}{n}} \right)$$

R = radius of the sphere, **n** = number of waves
m = number of sources – for symmetry reasons: 1 ≤ m ≤ n   n,m,k = 1,2,3,4,...
k = number of half basic wavelengths to reach a layer

n, m, k are integer parameters that provide by successively application, a table of layers **L**.

2.5.2.3 - Theorem:
With the radius **R** of the sphere all possible layers **L** are also given.

2.5.2.4 - Definition:

This simplified view of the **resulting field** is apparent:

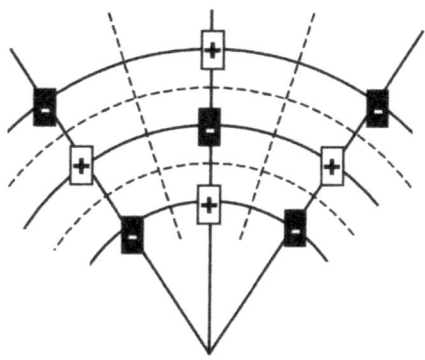

Illustration 2.5.3 – grid forming

solid lines = extremal lines = pole layers
dashed lines = zero lines = zero walls

There are **two** viewing options:

2.5.2.5 - Theorem: The **poles** form a grid-shaped radial **layer system**.

The pole layers are **stationary extremal states**.

If one looks at two pole layers lying on top of each other, these layers always own a counter phase order of their poles.

2.5.2.6 - Theorem: The **zero walls** form a grid-shaped **radial layer system**.

The **poles** are in the center of the each wrapping **zero field**.

2.5.3 – Normalisation

The layers on which the poles lie can be calculated after chapter 2.5 as follows:

$$L = R \cdot \left(\cos\frac{m\pi}{n} + \sqrt{4k^2 \cdot \sin^2\frac{\pi}{2n} - \sin^2\frac{m\pi}{n}} \right) \quad n,m,k = 1,2,3,4,\ldots$$

n, **k** and **m** are integer parameters that provide by successive application, a table of layers **L**. With the radius R of the sphere are all possible layers **L** given.

If **R = 1** is used a normalisation of the layers table is possible.

2.5.3.1 - Normalised stratification structure

n m	k	1	2	3	4	5	6	7	8
1 1		1	3	5	7	9	11	13	15
1 2		3	5	7	9	11	13	15	17
2 1		1	√7	√17	√31	7	√71	√97	√127
2 2		√2-1	2√2-1	3√2-1	4√2-1	5√2-1	6√2-1	7√2-1	8√-1
2 3		1	√7	√17	√31	7	√71	√97	√127
2 4		√2+1	2√2+1	3√2+1	4√2+1	5√2+1	6√2+1	7√2+1	8√2+1
3 3		0	1	2	3	4	5	6	7

A standardised stratification structure is depicted on the cover of this book. It is clear to see that the distribution is not uniform. Accumulation and also gaps are formed.

2.5.3.2 – General normalised stratification structure

n	m	L(k)
1	1	$2k-1$
1	2	$2k+1$
2	1	$\sqrt{(2k^2-1)}$
2	2	$k\sqrt{2}-1$
2	3	$\sqrt{(2k^2-1)}$
2	4	$1+k\sqrt{2}$
3	1	$½+\sqrt{(k^2 - ¾)}$
3	2	$-½+\sqrt{(k^2-¾)}$
3	3	$k-1$
4	1	$½\sqrt{2}+\sqrt{(k^2(2-\sqrt{2}) - ½)}$
4	2	$\sqrt{(k^2(2-\sqrt{2})-1)}$
4	3	$-½\sqrt{2}+\sqrt{(k^2(2-\sqrt{2})-½)}$
4	4	$k\sqrt{(2-\sqrt{2})}-1$
5	5	$½(\sqrt{5}-1)k-1$
6	1	$½\sqrt{3}+\sqrt{(k^2(2-\sqrt{3}) - ¼)}$
6	2	$½+\sqrt{(k^2(2-\sqrt{3})-¾)}$
6	3	$\sqrt{(k^2(2-\sqrt{3})-1)}$
6	6	$k\sqrt{(2--\sqrt{3})}-1$

2.6 – Radial stationary waves

The produced layers L f**orm a radial standing wave.**

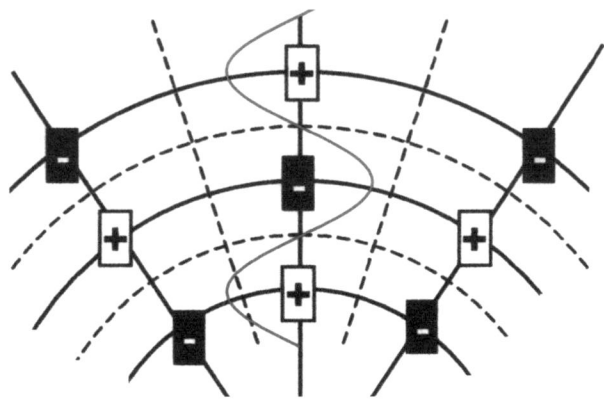

Illustration 2.6.1 – radial wave

The resulting layer frequencies and wavelengths can be calculated:

2.6.1 - Equation: $\lambda_n = L_{n+2} - L_n$

Each generated layer L can also be understood as **radial standing wave.**

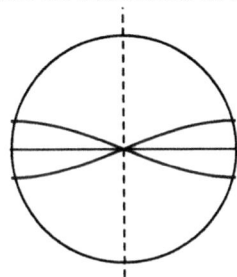

Illustration 2.6.2 – radial basic wave

The antinodes of oscillation so the oscillation maxima are always on the layer, as shown in Chapter 2.4 at the basic frequency.

To this basic oscillation suitable harmonics still come:

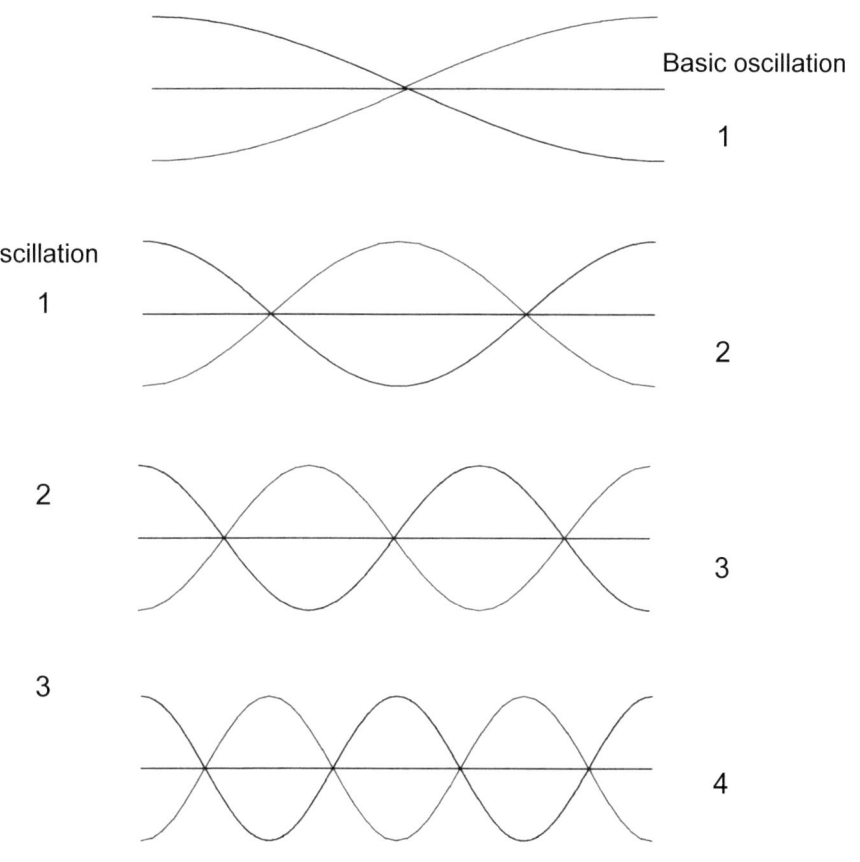

Illustration 2.6.3 – radial harmonic waves

2.6.2 - Equation: $\lambda_n = \dfrac{4L}{n}$ n = 1,2,3,4,...

2.6.3 - Theorem: The layers are **stationary** extremal states and form radial **standing** waves.

2.7 – Stratification structure

A standing wave on a sphere produces a **rotation-symmetric** spatial structure:

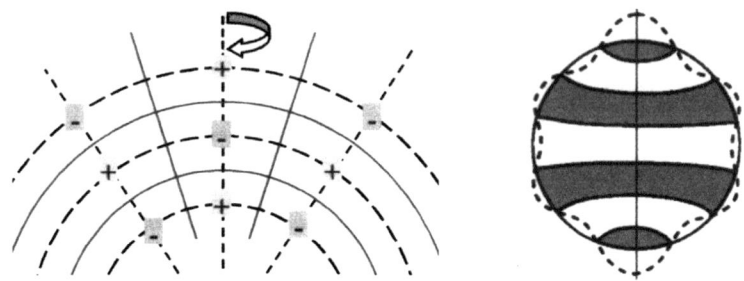

Illustration 2.7.1 – rotational symmetry

2.7.1 - Theorem: The **poles** are circular on **concentric** spheres.

2.7.2 - Theorem: The **zero areas** are concentric **cones** and **concentric spheres**.

The concentric cones from zero areas have an amazing similarity to configurations from the orbital model. (see 2.11.2)

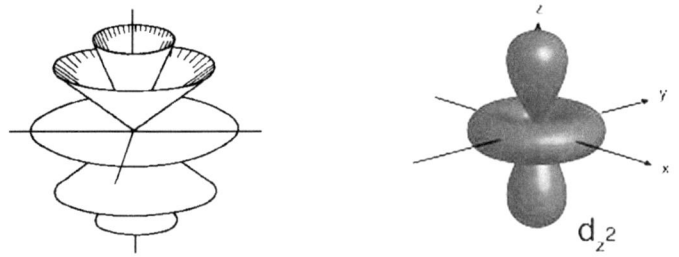

Illustration 2.7.2 – zero areas   Illustration 2.7.3 – atomic orbital

Together, these oscillation figures have even the property that they are **rotationally symmetric**.

2.7.1 - Definition: **Stratification structure**
= radial stratification structure generated by **one** standing wave

2.8 – Spatial grid

Two standing waves on a sphere create a radial grid-shaped structure:

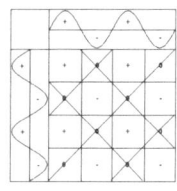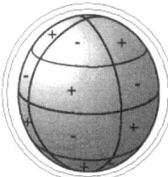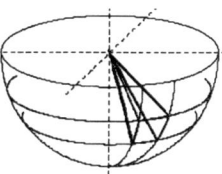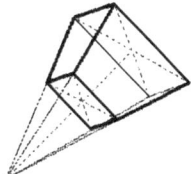

Illustration 2.8.1 – spatial grid

2.8.1 - Definition: Spatial grid
= **radial stratification system** generated by **two** standing waves.

Two viewing opportunities arise:

2.8.2 - Definition:

The **zero areas** form the walls of a grid-shaped radial oscillation system in the following **zero grid** called.

The **poles** are in the center of each **wrapping** zero cuboid.

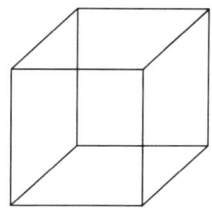

Illustration 2.8.2 – zero cuboid

2.8.3 - Definition:

The **poles** also form the walls of a grid-shaped radial oscillation system similar to a **molecular lattice** (NaCl) in the following **pole grid** called.

The **pole connections** behave like rods which swing freely on both ends.

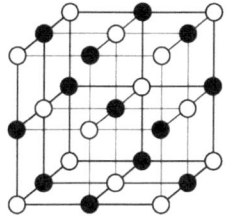

Illustration 2.8.3 – pole grid

2.9 – Spatial oscillation structure

Starting point is a standing wave around a ball

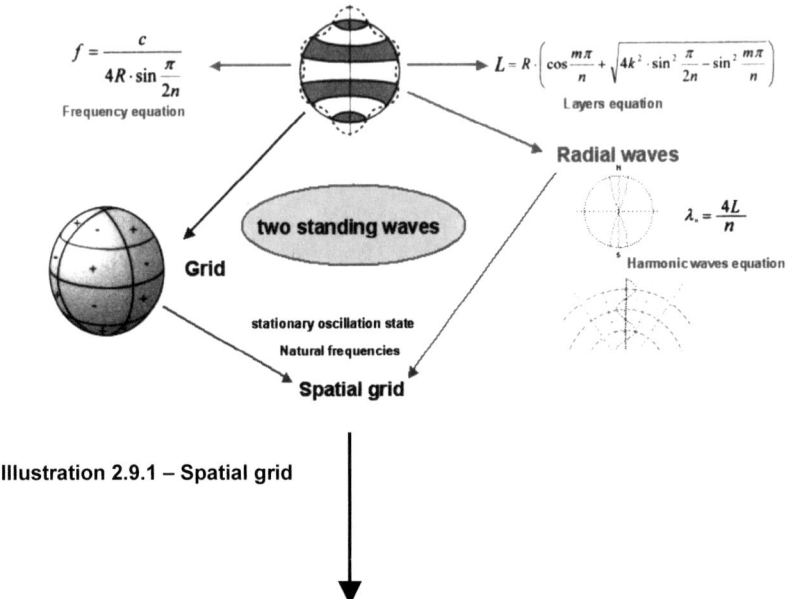

Illustration 2.9.1 – Spatial grid

2.9.1 - Definition: **Spatial oscillation structure**
= sum of all possible spatial grids on a sphere.

Consequence:

Only **1** standing wave is necessary to create a **spatial layering structure**.

2 Standing waves are necessary to create a **spatial oscillation structure**.

2.10 – Global net grids

With the definitions to the spatial grid and the oscillation structure you can introduce the notion of the global net grids. Generally, it can be defined as follows:

2.10.1 - Definition: Global net grid
   = Sum of spatial grids

The participating spatial grids form usually **harmonical** relations. This works for all rational numbers or numbers with fractional representation. The term harmony derives from the music and thinks the harmonious consonance of tones.

Example:

If an arbitrarily fundamental tone is chosen and is applied as 1, the third with 5:4 and the fifth with 3:2 results. All three tones then compose the sound combination known as the triad. Thus represent the numbers 1 and 5/4 and 3/2 as harmonical conditions.

2.10.2 - Theorem: Spatial oscillation structure
   = Sum of global net grids

Comment:

If grid by harmonical relationships stand with each other in relationship, there exist general reduction factors. Geometrically seen arise it from **overlapping** the grid walls or grid lines. It means that some grid walls and grid lines fall at least together.
Overlapping grids then belongs to the nature of the system.

2.11 – General attempt

With the construction of the spatial oscillation structure a mathematical and physical model is available that allows to explain the structures of the earth on a basis of waves.
The question is:
What is a general approach for an oscillation structure? The answer is found in the Laplace's equation.
Pierre-Simon (Marquis de) Laplace (28.03.1749 to 05.03.1827) was a French mathematician, physicist, and astronomer. He worked in the fields of probability theory and differential equations.
Laplace was always more physicists as a mathematician. He used the mathematics as a tool. Today, the mathematical procedures which Laplace developed and applied, become more important than his actual astronomical work.
The most important mathematical tools are the Laplace operator, Laplace's equation, Laplace's formula, as well as the Laplace transform.
The Laplace operator Δ is a mathematical operator that is a general mathematical provision (calculation way). The Laplace operator is a differential vector operator within the multidimensional analysis.
The Laplace operator occurs in Laplace's equation, for example. Twice continuously differentiable solutions of this equation are called harmonic functions.

Laplace's equation: $\quad \Delta f = 0$

Expressed in cartesian coordinates (x, y, z):

$$\Delta f = \frac{\partial^2 f}{\partial x^2} + \frac{\partial^2 f}{\partial y^2} + \frac{\partial^2 f}{\partial z^2} = 0$$

Applies in only one dimension:

$$\frac{d^2 f}{dx^2} = 0$$

This is the equation for a harmonic oscillator, such as a pendulum or a spring without friction.

> The Laplace's equation represents a mathematical formula to describe oscillation phenomena **in space**.

The common approach to a solution with a central configuration is to transform the Cartesian coordinates (x, y, z) in spherical coordinates (λ,φ,r).
Then, the entire function is decomposed into two part functions. Where a function the **radial part** contains and the other function the **part of the angle**.

The general approach to a solution function of Laplace's equation in spherical coordinates looks like this:

$$f(r,\lambda,\varphi) = \sum R(r) \cdot Y(\lambda,\varphi)$$

Both part functions **R** and **Y** can be solved in each case individually.

2.11.1 – Angle part

Following general solution function is specified for the part of the angle:

$$Y(\lambda,\varphi) = \frac{1}{\sqrt{2\pi}} \cdot N_m \cdot P_m \cdot \cos\lambda \cdot e^{im\varphi}$$

These are the spherical harmonics in complex notation.

Es gilt: $e^{im\varphi} = \cos(m\varphi) + i\cdot\sin(m\varphi)$

The N_m and P_m are the so called Legendre polynomials, which can be manipulated like constants in our consideration.

So results total:

$$Y(\lambda,\varphi) = \frac{1}{\sqrt{2\pi}} \cdot N_m \cdot P_m \cdot \cos\lambda \cdot (\cos m\varphi + i\cdot\sin m\varphi)$$

Multiplying the clips it is found:

$$Y(\lambda,\varphi) = \frac{1}{\sqrt{2\pi}} \cdot N_m \cdot P_m \cdot [\cos\lambda \cdot \cos m\varphi + i \cdot \cos\lambda \cdot \sin m\varphi]$$

And here you can see again the multiplicative associated sine and cosine functions, so tesseral spherical harmonics or grids. Here equipped with a real and an imaginary part, so a complex function as a general solution for the angle part of the Laplace's equation.

2.11.2 – Radial part

The radial component is widely considered as a possible solution:

$$R(r) = A \cdot e^{-kr} + B \cdot e^{kr}$$

Whereby here only one additive component as a solution may appear. What is dependent on the limiting conditions.
As yet be seen, obeys the earth oscillation structure with its radial part of Laplace's equation.

The radial equation applied on atomic configurations leads to the Schrödinger equation.
The behavior of quantum particles is described by the so-called wave function. The Wavefunction is obtained as a mathematical solution of the Schrödinger equation.
All wave equations together transport energy with the wave in one direction.
In a particle wave of the Schrödinger equation is transported not only energy, but also a particle, E.g. an electron. Like all other waves, a particle wave has two components that are specified mathematically as real and imaginary part of a complex function.
According to the Copenhagen interpretation of probability both parts of Schrödinger's equation together indicate the probability where the particle is. Where one or both parts are large, is the particle with probability to find. Where both parts are zero, the particle is guaranteed nonexistent. Thus, the solution of Schrödinger's equation provides probabilities for the stay of the particles. It develops the so called **orbitals**, so how they are used in chemistry.

Both the model of the atom, as well as the oscillation and layering structure developed in this book, use waves around a Center (see Chapter 2.0). So it should not surprise if there are matches in the oscillation figures like the illustrations 2.7.2 and 2.7.3 represented.

The resemblance is amazing, as even a number of physical quantities such as centrifugal potential, Coulomb potential and quantum-mechanical spin are used in the model of the atom.

These variables determine the limiting conditions of the underlying differential equation and so just create the characteristic figures of the orbitals.

2.11.3 – General remarks

A function that represents a solution of Laplace's equation has two properties:

1) It is twice continuously differentiable.

2) Its second derivative is equal to the origin function, multiplied by a constant.

Generally you can formulate for the radial part:

$$\frac{d^2 f}{dr^2} = k^2 \cdot f$$

This of course also applies to the angle part.

With the consequence that solution functions for the Laplace equation only **trigonometric functions** and **e-functions** are considered.

Part 2 – Applications 1

In part 1, a mathematical and physical basis was developed to explain the oscillation structures of the earth.

The developed model allows to calculate any kind of oscillation transmission, as long as they do not exceed the limits of the speed of light.

In the following, the model is applied on the geological layers and the layers of the atmosphere.

3.0 – Frequencies of the earth

According to Chapter 2.4 the fundamental frequency of an oscillation structure is:

$$f = \frac{c}{4R \cdot \sin\frac{\pi}{2n}} \quad n = 1,2,3,4,...$$

The fundamental frequency (n=1) for an oscillation structure is:

$$f_o = \frac{c}{4R}$$

The following values were used for the variable R, c:

c = 299792458 m/s as speed of light

The **WGS84** is a global geodetic reference system, that determine the basis of positions on the ground, and in the near earth space. The WGS84 delivers two earth radii:

Polar radius: 6356752 m
Equator radius: 6378137 m

There are two radii, thus also two frequencies:

n	Polar radius	Equator radius
1	11,7903 Hz	11,7508 Hz
2	16,6740 Hz	16,6181 Hz
3	23,5806 Hz	23,5016 Hz

3.0.1 - Theorem: The basic frequency of the Earth is about 11.7 Hz.
$$f_O \approx 11{,}7 \text{ Hz}$$

3.1 – Sferics

Between 1978 and 1979, special frequency measurements took place in Pfaffenhofen by Baumer and Sölling.
With a narrow bandwidth (2 kHz), the frequencies from 10 and 27 kHz were continuously recorded to 10 and 27 kHz by a reception facility. The range of the receiver was limited to 400-500 km.
This led to the discovery of the Sferics, which are also known as weather frequencies:

4150,84 Hz
6226,26 Hz
8301,26 Hz
10377,10 Hz
12452,52 Hz
28018,17 Hz
49810,08 Hz

See also" Das natürliche elektromagnetische Impuls-Spektrum der Atmosphäre" 1982 from Baumer and Eichmeier and in „Sferics" page 285, 1987 from Hans Baumer

The **Sferic basic frequency** is **4150,84 Hz**.

The other frequencies represent only harmonics, are therefore harmonic ratios to the sferic basic frequency.

A prime factorization with regard to the two basic frequencies of the earth is used as follows:

Equator radius $\quad$ 4150,84 : 11,75 = 353,263 ≈ 353 = prime

Polar radius $\quad$ 4150,84 : 11,79 = 352,064 ≈ 352

$$352 = 11 \cdot 32 = 11 \cdot 2^5$$

The back calculation for the Sferics proves:

For the Equator radius: $\quad$ $11{,}75 \cdot 11 \cdot 2^5$ = 4136,27 Hz

For the Polar radius: $\quad$ $11{,}79 \cdot 11 \cdot 2^5$ = 4150,19 Hz

The equatorial radius has a difference of 14.57 Hz to the sferic basic frequency.

The polar radius has a difference of 0.65 Hz to the sferic basic frequency.
The sferic basic frequency is used as a reference to define the earth basic frequency:

4150,84 : 352 = 11,79215909 = corrected basic frequency

3.1.1 - Definition: f_0 = **11,7921591 Hz =** Earth basic frequency

3.1.2 - Theorem: **The sferic basic frequency is the 5th octave of the 10th harmonic of the earth basic frequency.**

Comment:
11th natural frequency = 10th harmonic
Taking any frequency as basic frequency, then this is also known as first natural frequency. Then, the first harmonic to the basic frequency is the second natural frequency.
General: the n-th harmonic is the (n + 1)-th natural frequency.

Comparing the measured Sferic frequency with the two derived frequencies is the consequence that the Sferic frequency stands in relation to the **polar radius** and not to the equatorial radius.

The difference between pole frequency and the frequency of equator is about 14 Hertz. In his book "The cosmic octave" (pages 38-41), Cousto establishes a link between sferic frequency and sideral day.
After a sidereal day the Sun in relation to the stars is again in the same place in the sky. This corresponds to a geometric full rotation of the earth by 360° in a star fixed system. The mean sidereal day of the earth takes 23 hours, 56 minutes, 4,099 seconds.
The comparison of the sferics with the rotation of the earth provides a difference of about 3 Hz. To maintain the link between sferics and sideral day, Cousto devide down the sferic frequencies (page 199) and could then compare it with the devided generated tone scale.

Through this division, also the error is reduced accordingly. If one make a reference to the polar radius, this "calculating trick" is not necessary. It is also sometimes read, that the difference is interpreted as measurement error. Here should be in mind the following:
The measured value for the sferic basic frequency is specified with 2 digits behind the decimal point. Thus amounts the error not more than ±0.05 Hz or with a certain generosity ± 0.1 Hz.
If one is directed to the information of Cousto and interprets the measuring value with a mistake of 14 Hz, the whole measurement would be useless, because it would be afflicted with a systematic mistake.

Nevertheless, today's measurements, just in the electro technical field (also with frequency), are very precise, so that the information of 4,150.84 hertz is to be considered as correct.
As the oscillation approach also points, the sferic basic frequency can be derived (inaccuracy <0.7 Hz) with enough exactness.

The harmonics generated by the earth basic frequency are afflicted, about the whole frequency spectrum of the Sferics, with an inaccuracy of smaller than 0.7 Hz.
With Cousto the difference becomes bigger and bigger between sferic value and calculated equatorial value with rising frequency. Because of this lacking convergence this is, mathematically seen, a sure indicator for the fact that the solution published by Cousto shows only an approximation.

3.2 – Basic hull

The corrected basic frequency allows a back calculation for the earth's radius. Rearranging the equation for the basic frequency is:

$$f_o = \frac{c}{4R} \quad \longrightarrow \quad R = \frac{c}{4f_o}$$

The following values were used for the variable c, f_O:

f_O = 11,7921591 Hz = Earth basic frequency
c = 299792458 m/s as speed of light

The radius is:

3.2.1 - Definition: R = 6355758,426 m = L_0 = Basic hull

The comparison with the geodetic reference system WGS84 shows:

Polar radius: 6356752 m
Equator radius: 6378137 m

The basic hull, on which lie the zero lines and extremes (source points), is located at a radius of between **one and twenty kilometers below** the Earth's surface.

The oscillation structure for the earth can be defined:

3.2.2 - Definition: Earth oscillation structure
= Sum of **all** possible spatial grids on the basic hull.

3.3 – Table of the earth layers

By setting the basic hull L_0 in the equation for a layering structure one gets the equation for the **layers of the earth oscillation structure**.

3.3.1 - Equation:

$$L_{Erde} = L_o \cdot \left(\cos\frac{m\pi}{n} + \sqrt{4k^2 \cdot \sin^2\frac{\pi}{2n} - \sin^2\frac{m\pi}{n}} \right) \quad n,m,k \in \mathbb{N}.$$

The table of all layers of the earth results from this equation. Here, an excerpt up to n = 4:

k n m	1	2	3	4	5	6	7	8	
1 1	6355,76	19067,28	31778,79	44490,31	57201,83	69913,34	82624,86	95336,38	Km
2 1	6355,76	16815,76	26205,46	35387,37	44490,31	53554,57	62596,96	71625,76	Km
2 2	2632,64	11621,04	20609,44	29597,84	38586,24	47574,64	56563,04	65551,44	Km
2 4	15344,16	24332,56	33320,96	42309,36	51297,76	60286,16	69274,56	78262,96	Km
3 1	6355,76	14635,89	21433,41	27997,91	34476,36	40913,10	47326,39	53725,14	Km
4 1	6355,76	13122,94	18378,41	23426,02	28397,82	33333,04	38247,73	43149,72	Km
4 2		7365,95	13136,72	18390,65	23477,33	28486,50	33452,99	38393,37	Km
4 3		4134,54	9390,01	14437,62	19409,42	24344,64	29259,33	34161,32	Km
4 4		3373,22	8237,70	13102,19	17966,68	22831,16	27695,65	32560,14	Km

It must be still a reordering according to size. A **direct comparison** with the depths of the geological layers and the height of the atmospheric layers is possible with the sorted table.

3.4 – Analysis procedure

The concentric layers are not evenly distributed, but accumulate in some areas. For better optical evaluation is used, except the direct comparison, still the **mean layer method**.
The aim of this procedure is to determine the areas with oscillation accumulations and to get thus the maxima of the oscillation structure.

An example for the atmospheric layers:

From the **table of the earth layers** all values are searched out for n ≤ 20 which lie **beyond** the earth, up to a height of 700 km.

On 636 Km one receive 92 values ⇒ **6,9 Km per value = mean**

This mean value represents the mean rate of distribution of created layers by the oscillation structure.

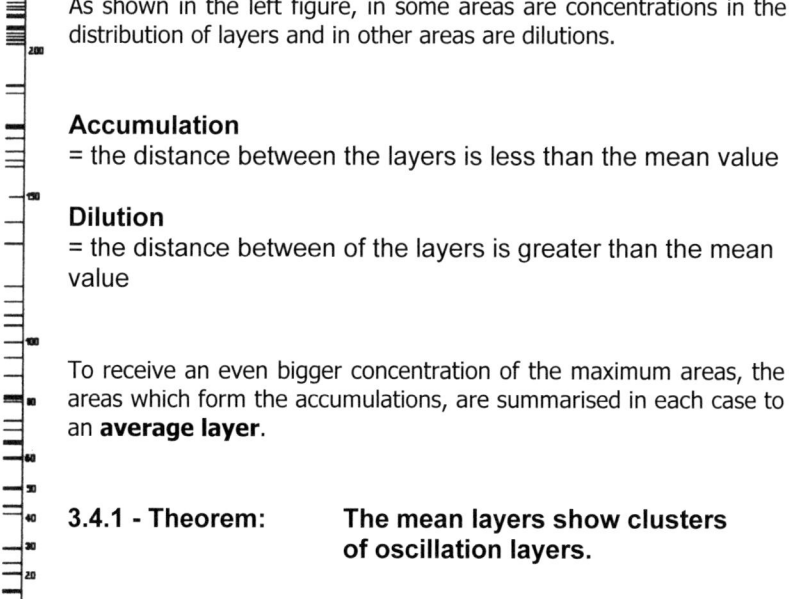

As shown in the left figure, in some areas are concentrations in the distribution of layers and in other areas are dilutions.

Accumulation
= the distance between the layers is less than the mean value

Dilution
= the distance between of the layers is greater than the mean value

To receive an even bigger concentration of the maximum areas, the areas which form the accumulations, are summarised in each case to an **average layer**.

3.4.1 - Theorem: **The mean layers show clusters of oscillation layers.**

3.5 – Geologic layers

From the **table of the earth layers** all values for n<11 are taken that lie **within** the earth.

There are 62 values at 6355,76 km   ⇒   102,5 Km per value

Then applies to the depth of the detected maxima layers:

3.5.1 - Equation: Depth = R_E – L

As a physical earth radius is used the mean radius R_E = 6371 Km, as it is common in physics.
The geological layers are compared with the values from the mean layer procedure. The following image 3.5.1 shows the layers for n<11:

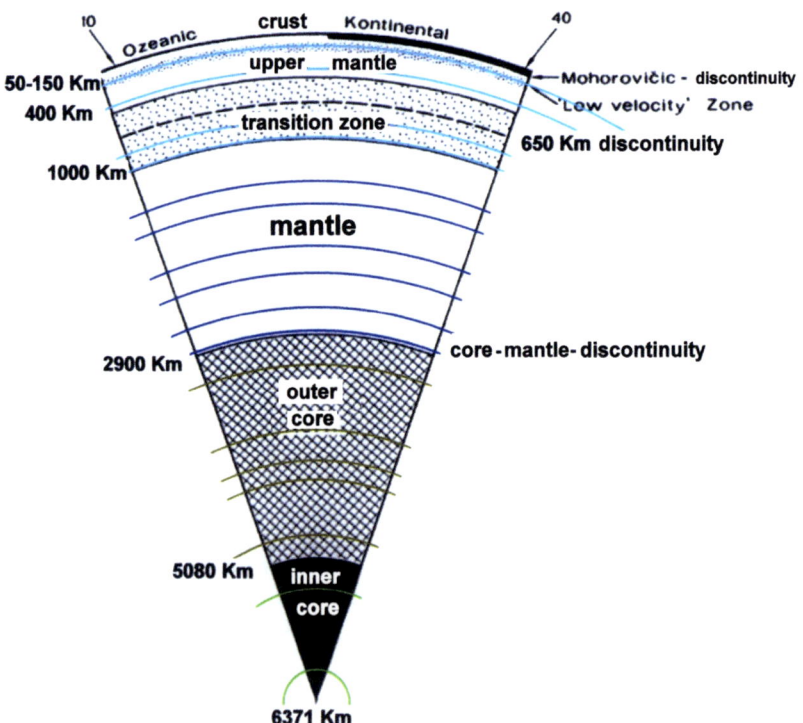

Illustration 3.5.1 – geologic layers for n<11

All values of the layer building, by the outer core to the surface, are contained in the earth layers structure. There are more calculated layers than geological layers.
A complete list of created layers can be found in „ Lattice structure of the Earth's magnetic field " chapter 13.

Greater accuracy is only achieved when increasing **n**, the number of oscillations. It is carried out evaluation for n<17.
The geological layers are compared with the values from the middle layers procedure again. The following image 3.5.2 shows the layers:

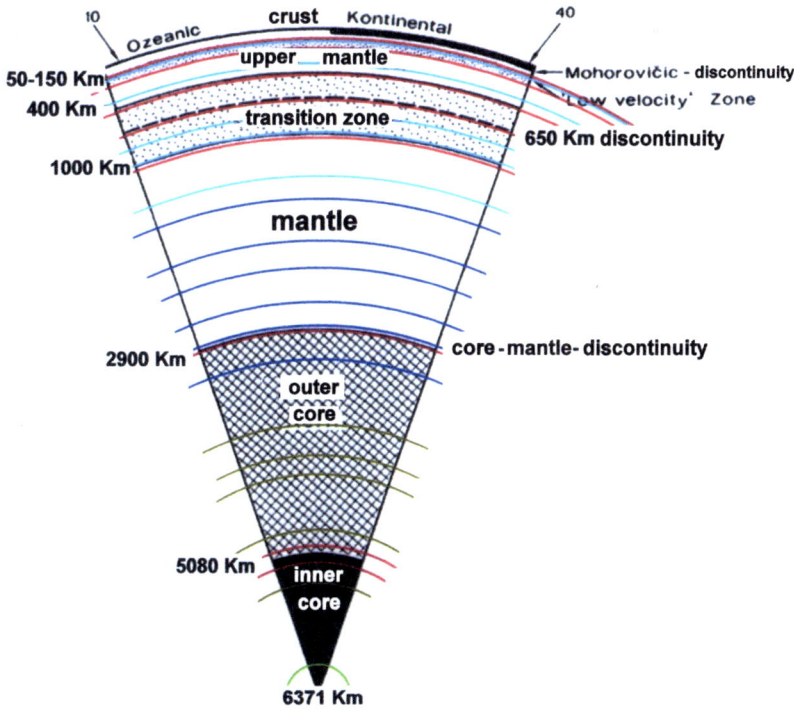

Illustration 3.5.2 – geologic layers for n<17

All values of the layer building are included in the earth layers structure. There are more calculated layers than geological layers.

3.5.2 - Theorem: The magnetic layers form the boundaries between **two** phases of matter.

With direct comparison of the calcuated layers with the geological layers arises a maximum difference of 14 kilometres for all layers, which corresponds to an inaccuracy of 0.22 per cent.
An exception is the core mantle discontinuity with 62 km difference, which corresponds to an inaccuracy by 1 percent.

Therefore applies:

3.5.3 - Theorem:

earth layer structure ⇔ earth oscillation structure

The earth layer structure is equivalent to the Earth oscillation structure.

Applies exactly:

3.5.4 - Theorem:

earth layer structure ⊂ earth oscillation structure

The earth layer structure is a subset of the earth oscillation structure.

3.5.5 - Implication: **The geological layers of the earth are representable as an oscillation phenomenon.**

Because the layers are at least 3 billion years old, the earth oscillation structure must have been also at that time.

The key question is therefore:

What is the **physical** connection between the model of the earth oscillation structure and the forming of the geological layers?

3.6 – Geologic layers and Laplace

Generally, you can carry together 16 relevant geological layers from the popular literature. The layers are sorted and numbered by depth.

n	Depth [Km]	Layer		ln(Depth)
	6371	center of the earth		
1	5100	border inner core / outer core		8,53699582
2	2900	border outer core / mantle		7,97246602
3	1700	1700 Km discontinuity		7,43838353
4	1200	1200 Km discontinuity		7,09007684
5	1000	border mantle / transition zone		6,90775528
6	920	920 Km discontinuity	900-1080 Km	6,82437367
7	720	720 Km discontinuity		6,57925121
8	660	660 Km discontinuity		6,49223984
9	520	520 Km discontinuity		6,25382881
10	410	border transition zone / upper mantle / 410km-disc.		6,01615716
11	300	X- discontinuity	250-350 Km	5,70378247
12	250	Lehmann discontinuity	190-250 Km	5,52146092
13	190	Lehmann discontinuity		5,24702407
14	100	low velocity zone		4,60517019
15	80	border upper mantle / Lithosphere		4,38202663
16	60	border Lithosphere / crust		4,09434456
	0	surface of the earth		

Right in the table is the logarithm naturalis (logarithm to the base e) for the respective depths.
The use of the logarithm happens because you can make an analysis easier. Function structures are more obvious.

The logarithm of the depth is represented as a function of numbering:

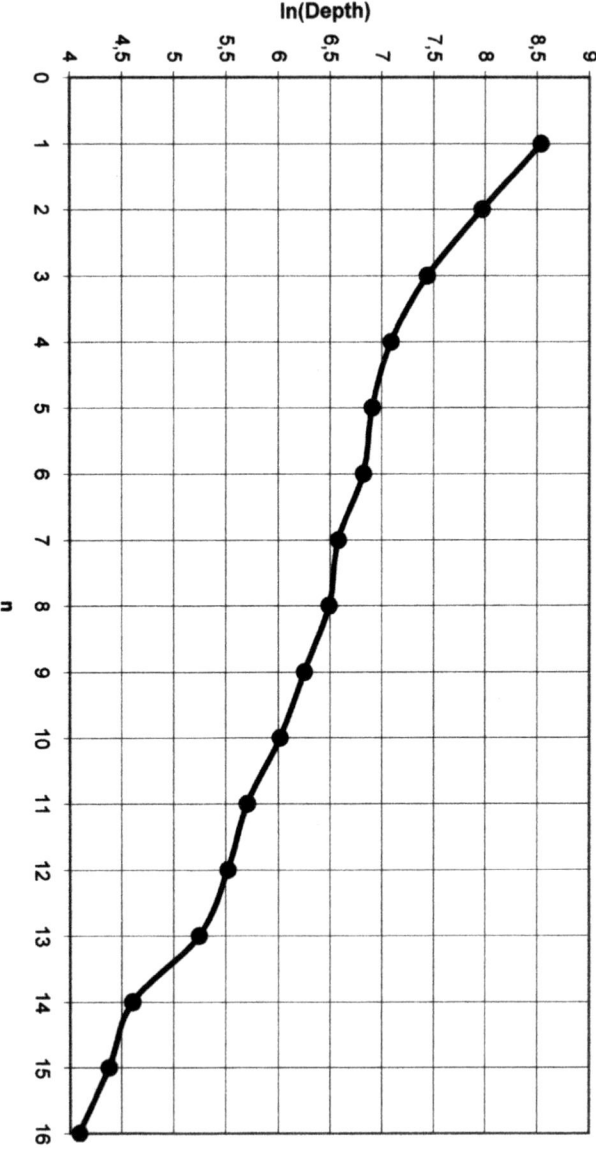

Illustration 3.6.1 – logarithm of the depth

The function in illustration 3.6.1, looks first time only some parts are approximately linear. If you look but closer to the course so you can see:

a) between point 8 and 13, the slope is nearly constant
b) between points 1 and 2 as well as between points 2 and 3 and between thee points 13 and 14, the incline is so great that there still a point can be inserted to flatten the slope.
c) between point 5 and 6 and 7 and 8, the incline is so small that there the numbering can be used on a half, increasing so the slope.

For reasons of practical mathematical handling regarding the function to be determined, it is better to start numbering with zero. The layers that are corrected in the numbering arranged by depth, then results in the following table:

n	Depth [Km]	Layer		ln(Depth)
	6371	center of the earth		
0	5100	border inner core / outer core		8,53699582
1				
2	2900	border outer core / mantle		7,97246602
3				
4	1700	1700 Km discontinuity		7,43838353
5	1200	1200 Km discontinuity		7,09007684
5,5	1000	border mantle / transition zone		6,90775528
6	920	920 Km discontinuity	900-1080 Km	6,82437367
7	720	720 Km discontinuity		6,57925121
7,5	660	660 Km discontinuity		6,49223984
8	520	520 Km discontinuity		6,25382881
9	410	border transition zone / upper mantle / 410km-disc.		6,01615716
10	300	X- discontinuity	250-350 Km	5,70378247
11	250	Lehmann discontinuity	190-250 Km	5,52146092
12	190	Lehmann discontinuity		5,24702407
13				
14	100	low velocity zone		4,60517019
15	80	border upper mantle / Lithosphere		4,38202663
16	60	border Lithosphere / crust		4,09434456
	0	surface of the earth		

The newly added layers in the numbering of 1, 3 and 13 are clear to see in the table. The corrected function looks like this:

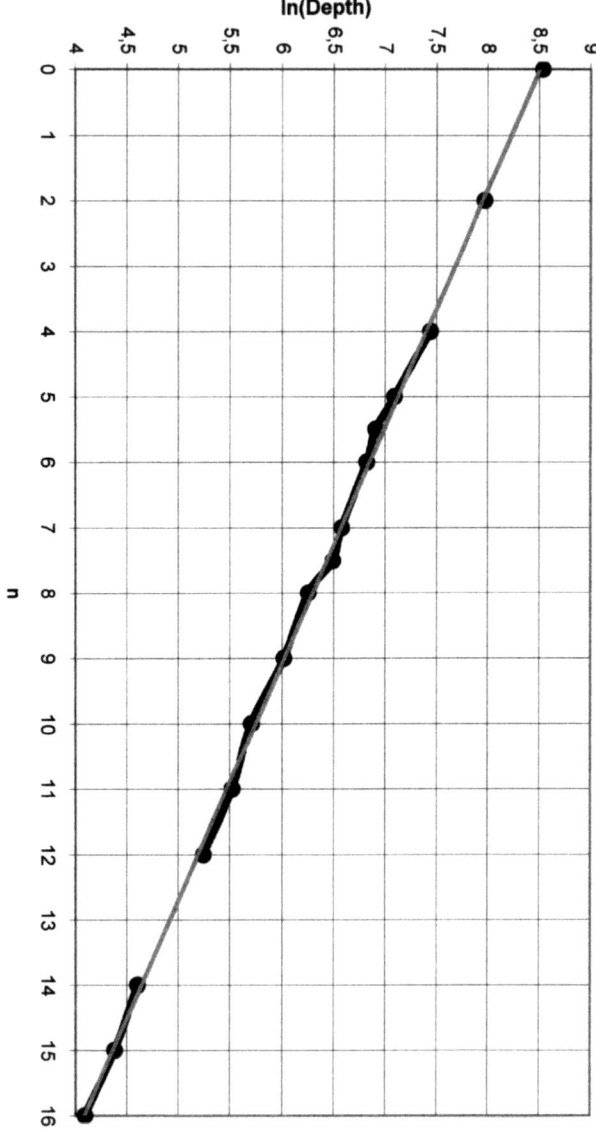

Illustration 3.6.2 – logarithm of the depth

The gray line in the illustration 3.7.2 represents a linear function that was obtained by linear regression from the corrected table. It is to see that the layer values match well with the approximation function.
So it can be used here a linear function for the geological layers as solution approach.

It arises: **ln(Depth) = - 0,2777·x + 8,537**

It is obtained by rearranging:

3.6.1 - Equation: Depth = 5100·e$^{-0,2777·x}$

Putting **x=n** then following function arises for equation 3.6.1:

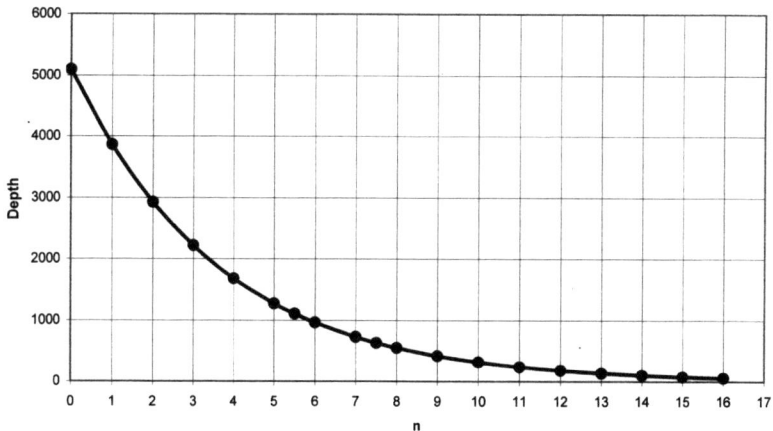

Illustration 3.6.3 – depth of the geologic layers

Equation 3.6.1 has all of the properties that are necessary to 2.11.3, to come as a function of solution of the Laplace equation into account. Thus the geological layers represent a solution of Laplace's equation, specially of the radial component.
In consequence this theorem can be set up:

3.6.2 - Theorem: The geological layers are an expression of an oscillation phenomenon.

On the equation 3.6.1 can be still made simplifications.

It applies: $\qquad 0{,}2777 = 3{,}6^{-1} = 8/15$

When all values are used:

3.6.3 - Equation: $\qquad T_n = 5100 \cdot e^{\frac{-8n}{15}} \qquad$ [Km]

It can be made the following relationship:

$r_{ik} = R_E/5$ $\qquad$ **r_{ik} = inner core** and $R_E = 6371$ Km

and it still applies:

$5100 = R_E - r_{ik} = 4/5\ R_E = 4 r_{ik}$

Then, generally you can write:

3.6.4 - Equation: $\qquad \boxed{T_n = 4 \cdot r_{ik} \cdot e^{\frac{-8n}{15}}} \qquad$ [Km]

Then, you can write for the geological layers:

3.6.5 - Equation: $\qquad \boxed{T_n = \frac{4}{5} \cdot R_E \cdot e^{\frac{-8n}{15}}} \qquad$ [Km]

The layers arranged by depth and the calculated values result in the following table:

n	Depth [Km]	Layer	Calculated Depth [Km]
0	5100	border inner core / outer core	5100
1			**3863,505**
2	2900	border outer core / mantle	2926,798
3			**2217,196**
4	1700	1700 Km discontinuity	1679,637
5	1200	1200 Km discontinuity	1272,409
5,5	1000	border mantle / transition zone	
6	920	920 Km discontinuity 900-1080 Km	963,913
7	720	720 Km discontinuity	730,212
7,5	660	660 Km discontinuity	
8	520	520 Km discontinuity	553,172
9	410	border transition zone / upper mantle / 410km-disc.	419,055
10	300	X- discontinuity 250-350 Km	317,455
11	250	Lehmann discontinuity 190-250 Km	240,488
12	190	Lehmann discontinuity	182,182
13			**138,012**
14	100	low velocity zone	104,551
15	80	border upper mantle / Lithosphere	79,202
16	60	border Lithosphere / crust	60

The mean error of the calculated values for the geological layers are below 1 per cent.

In addition, three layers are created.

3.7 – Layers of the atmosphere

From the **table of the earth layers** all values are used for n<20, which are **outside** the earth until 640 km height.

There are 104 values at 639 km   ⇒   6,14 Km per value

Then applies to the height of the detected maxima layers:

3.7.1 - Equation: Height = L - R_E

As a physical earth radius is used the mean radius R_E = 6371 Km, as it is common in physics.
The layers of the atmosphere are compared directly with all calculated layers and then with the mean layer procedure.

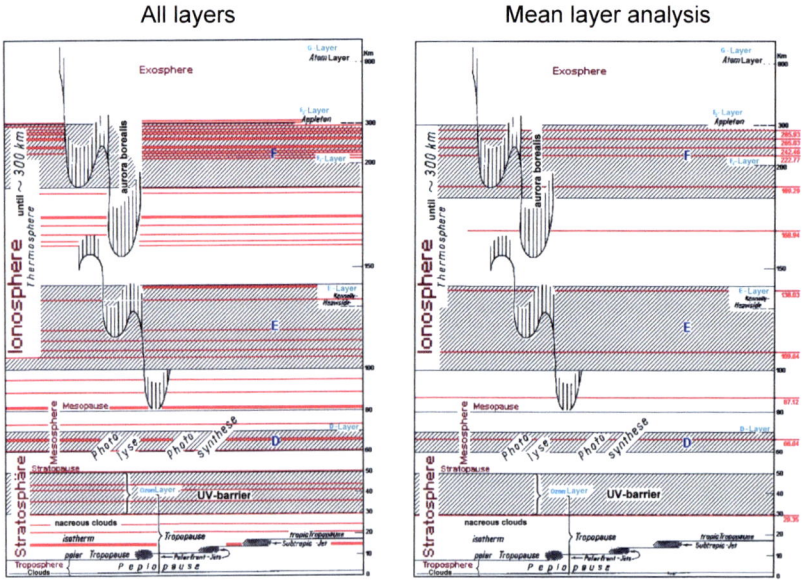

Illustration 3.7.1 – atmospheric layers

All generated layers used by the mean layer procedure lie in the areas of atmospheric layers, except for two layers. So the **earth layers structure** includes **all** layers of the atmosphere to a height of 300 km.

The ozon, D, E, F - layers are representable as an oscillation phenomenon. Oscillation clusters arise also in the gaps between the layers.

3.7.2 - Implication: **The atmosphere can be interpreted as an oscillation phenomenon.**

3.7.3 - Evaluation of the atmospheric layers

Layer	Height [km]	Center distance [km]	Layer [km]	n	m	k
Ozon	30	6401	6401,658	18	2	2
Ozon	50	6421	6421,979	15	2	2
D	60	6431	6431,846	14	2	2
D	70	6441	6444,102	13	2	2
E	100	6471	6477,747	18	3	3
E			6479,551	11	2	2
E	140	6511	6510,603	20	13	11
E			6510,584	16	3	3
E			6505,909	10	2	2
F	180	6551	6558,809	14	3	3
F			6546,103	11	7	6
F			6545,820	17	17	11
F			6541,736	9	2	2
F	300	6671	6674,749	13	8	7
F			6666,665	7	2	2

Direct comparison for **m=K=2** respectively a layer exists. That you can use for a simplification of the layer equation.

For **m=K=2** the atmospheric layers, relative to the center of the earth, **are directly derivable from the earth layers structure**. It is then:

3.7.4 - Equation:

$$L = L_o \cdot \left(\cos\frac{2\pi}{n} + \sqrt{16 \cdot \sin^2\frac{\pi}{2n} - \sin^2\frac{2\pi}{n}} \right)$$

Atmospheric Equation

3.7.5 - Mean layers for n≤20

n	Center distance [km]	Height [km]
1	31778,792	25407
2	11621,041	5250
3	8280,127	1909
4	7365,952	994,9
5	6982,090	611,1
6	6783,318	412,3
7	6666,665	295,6
8	6592,213	221,2
9	6541,736	170,7
10	6505,909	134,9
11	6479,551	108,5
12	6459,588	88,5
13	6444,102	73,1
14	6431,846	60,8
15	6421,979	50,9
16	6413,917	42,9
17	6407,244	36,2
18	6401,658	30,6
19	6396,936	25,9
20	6392,907	21,9

The **atmospheric equation** provides for **6<n<21** a total of 13 layers of up to 300 km height. If one put these layers in a drawing for the atmosphere, the following picture arises.

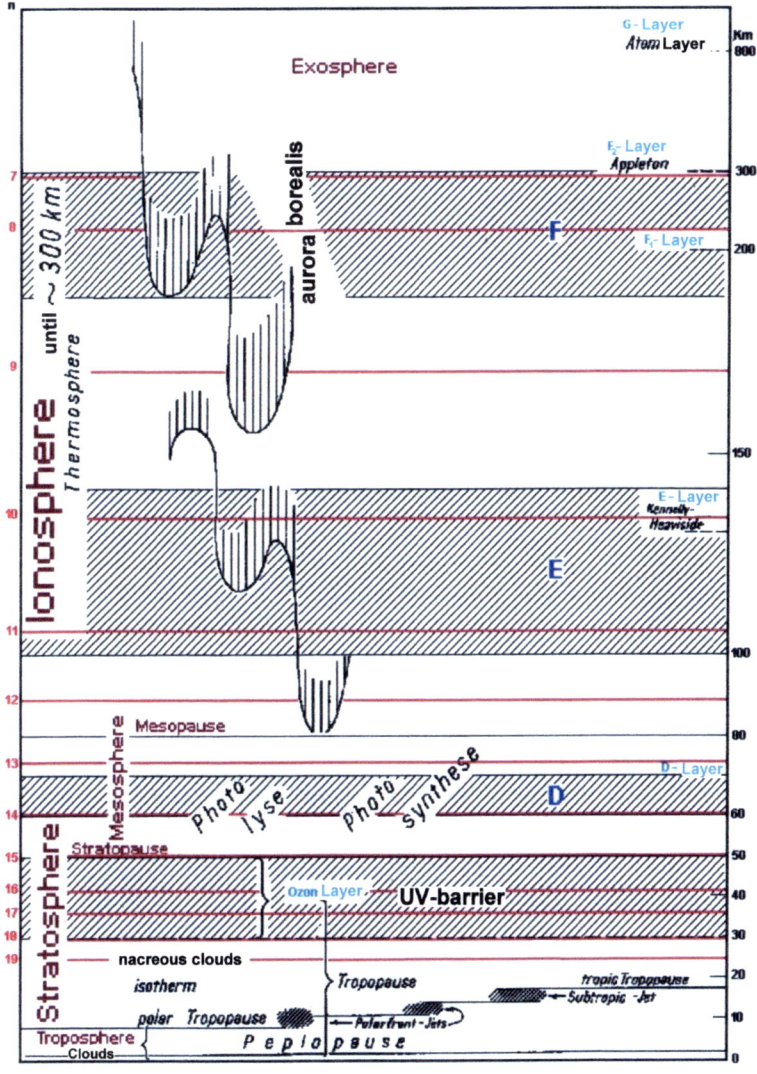

Illustration 3.7.2 – atmospheric layers

3.7.6 - Implication: The atmosphere is representable as an oscillation phenomenon.

It applies:

3.7.7 - Theorem:

atmospheric layers $\Leftrightarrow$ earth oscillation structure

The atmospheric layers of Earth are equivalent to the earth oscillation structure.

Applies exactly:

3.7.8 - Theorem:

atmospheric layers of the earth $\subset$ earth oscillation structure

The atmospheric layers of Earth are a subset of the earth oscillation structure.

The ozone, D, E and F layers also form the electrically conducting layers of the atmosphere. And, thanks to their potential structure, they are significantly involved in the electric field of the earth. Atmospheric layers and the electric field of the earth are coupled together.

So the question is:

What is the **physical** relation between the model of the oscillation structure of the earth and the forming of the atmospheric layers or of the electric field of the earth?

3.8 – Layers of the atmosphere and Laplace

Generally, you can put 11 relevant atmospheric layers together. The individual layers are sorted and numbered according to height.

n	Height		ln(Height)
0	20	Tropopause	2,99573227
1	30	Ozon	3,40119738
2	50	Ozon	3,91202301
3	60	d	4,09434456
4	70	d	4,24849524
5	100	e	4,60517019
6	140	e	4,94164242
7	180	f1	5,19295685
8	200	f1	5,29831737
9	300	f2	5,70378247
10	800	g	6,68461173

Right in the table is the logarithm naturalis for the respective heights.

The logarithm of height is represented as a function of numbering:

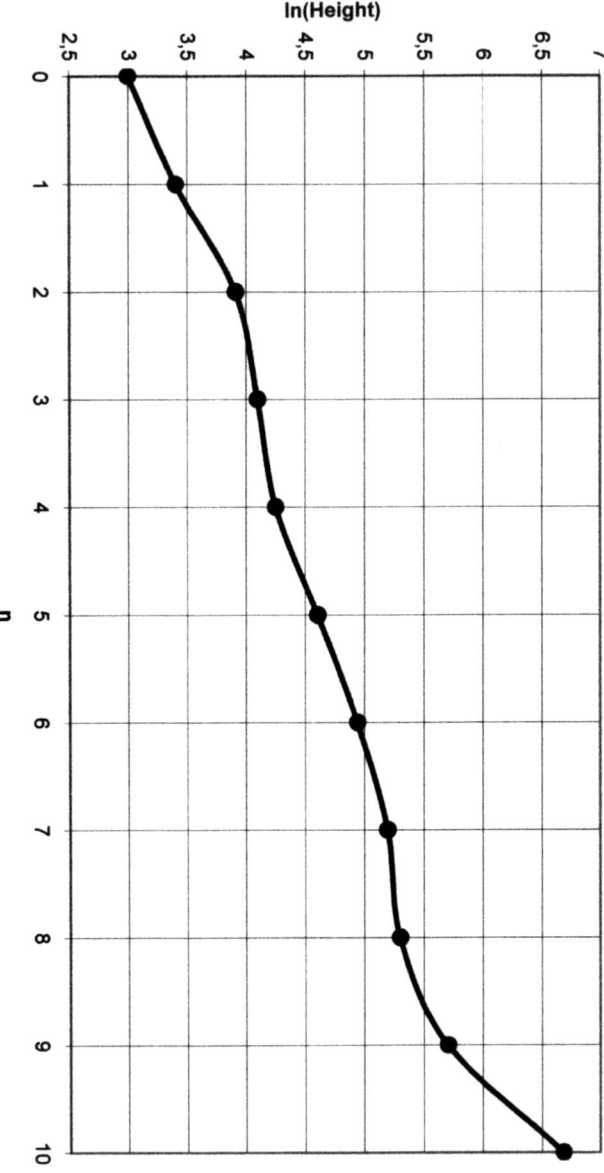

Illustration 3.8.1 – logarithm of height

The function in illustration 3.8.1, seems to be not very linear on first view except in few parts. If you look but closer to the course so you can see:

a) Between the points 2 to 7, the slope is nearly constant.
b) Between points 1 and 2, as well as between points 8 and 9 is the slope so large, there can still a point be inserted, to flatten the slope.
c) Between points 9 and 10 the slope so large that there still several points can be inserted. Therefore, point 10 is eliminated first.
d) Between point 7 and 8, the slope is so small that there the numbering can be used on a half, increasing so the slope.

The following table shows the layers that are corrected in the numbering arranged by height:

n	Height		ln(Height)
	[km]		
0	20	Tropopause	2,99573227
1	30	Ozon	3,40119738
2			
3	50	Ozon	3,91202301
4	60	d	4,09434456
5	70	d	4,24849524
6	100	e	4,60517019
7	140	e	4,94164242
8	180	f1	5,19295685
8,5	200	f1	5,29831737
9			
10	300	f2	5,70378247

The newly added layers at the numbers 2 and 9 are clear to see in the table. The corrected function looks like this:

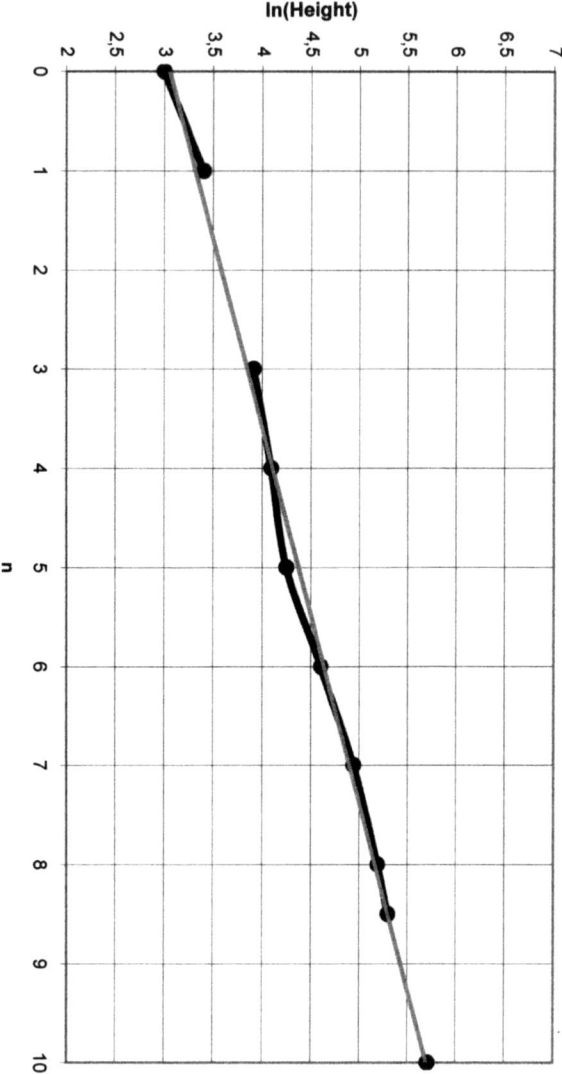

Illustration 3.8.2 – logarithm of height

The gray line in illustration 3.8.2 represents a linear function that was obtained by linear regression from the corrected table. Like is to see the layers values match well with the approximation function.
It can be used here so a linear function for the atmospheric layers as a solution.

It is as follows: **ln(Height) = 0,277·x + 2,9957**

It is obtained by rearranging:

3.8.1 - Equation: Height = 20·e$^{0,277·x}$

Putting x=n then following function arises for equation 3.8.1:

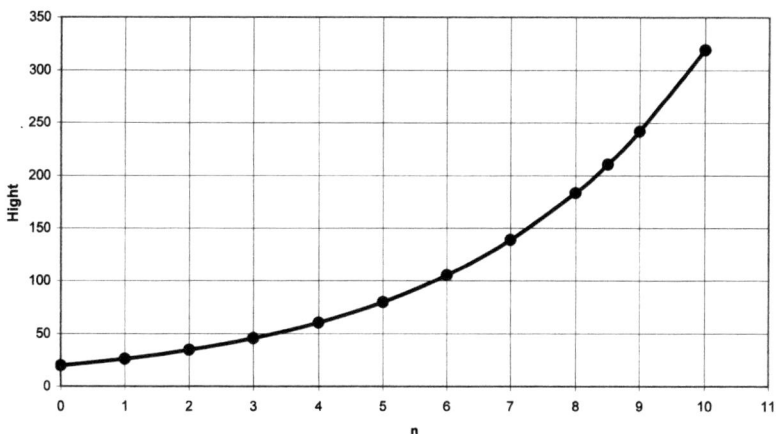

Illustration 3.8.3 – Height

Equation 3.8.1 has all the properties that are necessary to get as a solution function of Laplace's equation in consideration to 2.11.3. Thus the atmospheric layers represent a solution of Laplace's equation, specially for the radial part.
In consequence the following sentence can be set up:

3.8.2 - Theorem: The atmospheric layers are an expression of an oscillation phenomenon.

On the equation of 3.8.1 can be still made simplifications.

It still applies: $\quad 0{,}277 = 3{,}6^{-1} = 8/15$

When all values are used:

3.8.2 - Equation: $\quad H_n = 20 \cdot e^{\frac{8n}{15}} \quad$ [Km]

It can be made to the following relation:

$r_{ik} = R_E/5 \qquad r_{ik} =$ **inner core** and $R_E = 6371$ Km

and it still applies:

$5100 = R_E - r_{ik} = 4/5\, R_E = 4 r_{ik}$

$5100 = 255 \cdot 20 \qquad => \qquad 20 = 4/255 \cdot r_{ik}$

Then, you can generally write:

3.8.3 - Equation: $\quad \boxed{H_n = \frac{4}{255} \cdot r_{ik} \cdot e^{\frac{8n}{15}}} \quad$ [Km]

3.8.4 - Equation: $\quad \boxed{H_n = \frac{4}{1275} \cdot R_E \cdot e^{\frac{8n}{15}}} \quad$ [Km]

The layers arranged by height and the calculated values result in the following table:

n	Height		Calculated Height
	[Km]		[Km]
0	20	Tropopause	20
1	30	Ozon	26,383
2			**34,803**
3	50	Ozon	45,912
4	60	d	60,565
5	70	d	79,896
6	100	e	105,396
7	140	e	139,035
8	180	f1	183,411
8,5	200	f1	210,657
9			**241,950**
10	300	f2	319,172

The mean error of the calculated values for the layers is below 2 per cent.

In addition, even two layers arise.

3.9 – Planetary oscillation systems

The earth with their geological layers and the atmosphere layers represent solutions of the radial component of Laplace's equation, resulting in oscillation phenomena.

It can be generally written for the geological layers:

$$T_n = 4 \cdot r_{ik} \cdot e^{\frac{-8n}{15}}$$

It can be generally written for the atmospheric layers:

$$H_n = \frac{4}{255} \cdot r_{ik} \cdot e^{\frac{8n}{15}}$$

So it be:

k=0   for atmospheric layers
k=1   for geological layers

Then applies to all layers of the earth:

3.9.1 - Equation:
$$H_{k,n} = 255^{k-1} \cdot 4 r_{ik} \cdot e^{(-1)^k \cdot \frac{8n}{15}} \quad [Km]$$

with $r_{ik}=R_E/5$ and r_{ik} = **inner core** and R_E= 6371 Km

you can also write:

3.9.2 - Equation:
$$H_{k,n} = 255^{k-1} \cdot \frac{4}{5} R_E \cdot e^{(-1)^k \cdot \frac{8n}{15}} \quad [Km]$$

For the geologic and the atmospheric layers it was shown in two things way that it is an oscillation phenomena:

a) direct comparison with calculated layers
b) as a solution of the radial part of Laplace's equation

The geological and the atmosphere layers are therefore **planetary systems with an oscillation structure**.

3.9.3 - Definition: Planetary oscillation system
= Planetary system
with an oscillation structure

3.9.4 - Theorem: The structure of the two planetary oscillation systems, the geological layers and the atmospheric layers, are represented by **a single** oscillation structure.

3.9.5 - Theorem:

Earth layer $\subset$ Laplace solution $\subset$ Earth oscillation structure

Generally the following statement can be generated:

3.9.6 - Theorem: **Stratification around or in a central solid is always the expression of an oscillation phenomenon.**

Geological layers, atmospheric layers and also electrons on their orbitals around the nuclei obey the same laws of oscillation, resulting from Laplace's equation. Here the question originates whether this oscillation principle is, much more of a general nature and arises here as a structural principle of the universe?

3.10 – Layers and frequencies

According to chapter 2.6 a frequency can be associated with any layer. The basic oscillation behaves as if the pole diameter of the earth would swing freely at its poles.

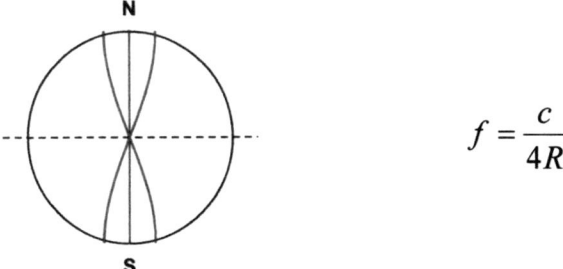

$$f = \frac{c}{4R}$$

Illustration 3.10.1 – basic oscillation

From the table of the earth layers, you can generate such a **frequency table**. For **n = 1** results:

k	1	2	3	4	5	6	7	8	
n m									
1 1	11,79	3,93	2,358	1,68	1,31	1,07	0,907	0,786	Hz
	6355,76	19067,28	31778,79	44490,31	57201,83	69913,34	82624,86	95336,38	Km

For **n = 1, k = 2** results:
3,93 Hz = the half Schumann frequency = $f_S /2$

For **n = 1** yet another easy relation between the basic hull and the other generated layers exist:

	k	1	2	3	4	5	6	7	8
n	m								
1	1	L_o	$3L_o$	$5L_o$	$7L_o$	$9L_o$	$11L_o$	$13L_o$	$15L_o$

This results in the following context:

3.10.1 - Equation:

$$\frac{f_S}{2} = \frac{f_o}{3} \longrightarrow f_S = \frac{2}{3} f_o$$

The earth frequency is one fifth of the Schumann frequency

According to chapter 2.6 applies to the first harmonic:

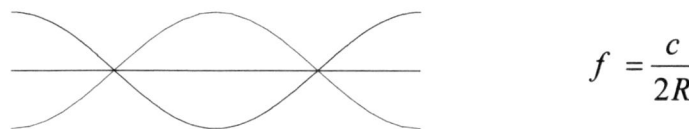

$$f = \frac{c}{2R}$$

Illustration 3.10.2 – harmonic oscillation

k	1	2	3	4	5	6	7	8	
n m									
1 1	23,584	7,861	4,716	3,369	2,62	2,143	1,814	1,572	Hz
	6355,76	19067,28	31778,79	44490,31	57201,83	69913,34	82624,86	95336,38	Km

3.10.2 - Theorem:
The Schumann frequency is part of the frequency spectrum of the earth.

3.11 – Schumann frequency

The German physicist Dr. W. O. Schumann from the Technical University of Munich in 1952 made exercises for the electricity with his students. Theme at that time was the calculation of **cavity resonators**.

While pretending that the outer ball of the resonator should be the base of the ionosphere, so the heaviside layer, the inner ball should be the surface of the earth. The question was to determine the natural frequency (resonance) of this cavity resonator. As a result, 7.8 Hz was received.

This task can be solved only by use of the differential and integral calculus. And the solution is with the discovery of so-called transversal electromagnetic waves (TM) in a cavity resonator. These are now known as **Schumann waves** or also **Schumann resonances**.

The Schumann frequency is the natural frequency of the earth surface - ionospheric cavity resonator. The consequence is:

3.11.1 - Theorem   The Schumann frequency is the
                        natural frequency of the atmosphere.

3.11.2 - Definition: Schumann frequency = f_S = 7,83 Hz

Two simple relations of earth frequency and the Schumann frequency to the basic hull L_0 are derived from equations 2.4.3 and 3.10.1:

3.11.3 - Equations:

$$f_O = \frac{c}{4L_O} \qquad\qquad f_S = \frac{c}{6L_O}$$

Yet the following relation to the sferics exist:

Sferic basic frequency: 4150,84 Hz = $11 \cdot 2^5 f_O$ and f_O = 3/2 f_S

Also it is therefore:

Sferic basic frequency: $4150{,}84\ Hz = 33 \cdot 2^4\ f_s$
33. natural wave = 32. harmonic wave

3.11.3 - Theorem: **The sferic basic frequency is the 4th octave of the 32. harmonic of the Schumann frequency.**

3.12 – Summary

The Sferics are proved physically as electromagnetic waves.

The sferic basic frequency is the 5th octave of the 10 harmonic of the earth frequency. The Sferic basic frequency is the 4th octave of the 32. harmonic of the Schumann frequency at the same time.

The sferic frequencies are included in the spectrum of the earth frequencies.

The Schumann frequency is proved physically as an electromagnetic wave.

The Schumann frequency is two-thirds the rate of earth frequency – the Earth frequency is the **fifth** of the Schumann frequency.

The Schumann frequency is included in the spectrum of the earth frequencies.

There is a connection between electromagnetic oscillations and the earth oscillation structure.

Here arises the question: is there a connection of the earth's magnetic field or electric field with the earth oscillations structure? To verify this, the Earth magnetic field must be taken into consideration.

Part 3 – Applications 2

With the construction of the oscillation structure in part 1 is a mathematical and physical model available, that allows to explain the concentric structures such as layer building of the earth, based on an oscillation structure.

With the applications in part 2, the model was applicated on the geological layers and the layers of the atmosphere.

As a result of the statements and the question in Chapter 3.12, the applicated model is now on the electromagnetic field of the earth.

4.0 – Earth magnetic field

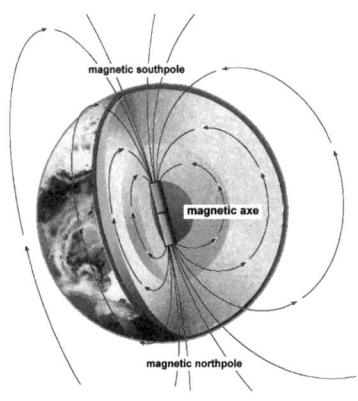

Illustration 4.0.1 – dipole field

From the school and from the media, we always know the earth's magnetic field as a field that corresponds to the field of a bar magnet. It is due to the so called dipole field.

Historical this view of the earth's magnetic field explains the behavior of a needle of inclination.

It is available at the pole perpendicular to the earth's surface and at the equator level parallel to the surface of the earth.

Taking measured values of the field, ít can be shown graphically that four magnetic poles (Illustration 3.4.2) exist and that the dipole model is not sufficient to explain exactly the real earth's magnetic field.

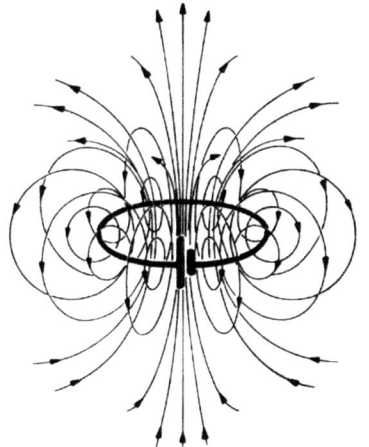

Illustration 4.0.2 – current loop

The physical approach for such a dipole field is the contemplation of the magnetic field of a so called **current loop**.

The mathematical derivation leads to a differential equation in which a so-called **elliptic integral** occurs, for which no closed mathematical solution - in the form of an equation - exists.

The mathematically commonly used approach is **converting** the appearing term in the integral in an **infinite sequence:**

Simplified can be written:

$$a_1 \cdot x + a_2 \cdot x^2 + a_3 \cdot x^3 + a_4 \cdot x^4 + ...$$

Then just **cut off this sequences after the first link**. Now integrate the rest, so the general equation for the dipole field, which only depends on the geographical latitude φ is created.

4.0.1 - Equation:

$$B = \frac{\mu_0 \cdot m}{4 \cdot \pi \cdot r^3} \cdot \sqrt{1 + 3 \cdot \cos^2(90 - \varphi)} \qquad \mu T$$

The mathematical approach presented here to the achievement of the dipole equation can be considered due to the cut off limbs of the rest, merely as a **first approximation**.

If you start to integrate the remaining limbs of the infinite sequence (from the current loop consideration), so you get the quadrupole field, the Octupole field, etc. Overall, this is called **multi pole forming**.

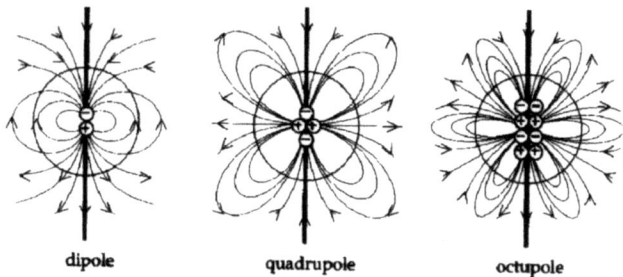

Illustration 4.0.3 – multipole forming

4.1 – Gauß and Weber

Already Gauß and Weber recognized in 1838, in their experiments with the earth's magnetic field, that the magnetic field can not be explained just by the model of a rod magnet or a current loop.

In 1838 appeared the "general theory of terrestrial magnetism" of C.F. Gauß and W. Weber where they specify the following potential equation for the magnetic field of the earth:

4.1.1 - Equation:

$$V(\varphi, \lambda, r, t) = a \cdot \sum_{n=1}^{N} \sum_{m=0}^{n} \left(g_n^m(t) \cdot \cos(m\lambda) + h_n^m(t) \cdot \sin(m\lambda) \right) \cdot \left(\frac{a}{r} \right)^{n+1} \cdot P_n^m \cdot \sin\varphi$$

The magnetic flux density **B** can be derived as a vector through the gradient from the potential equation:

4.1.2 - Equation:

$$B(\varphi, \lambda, r, t) = -\nabla V(\varphi, \lambda, r, t)$$

According to equation 4.1.2 the magnetic flux density B is depending on the latitude φ, the longitude λ, distance **r** from the center and the time **t**.

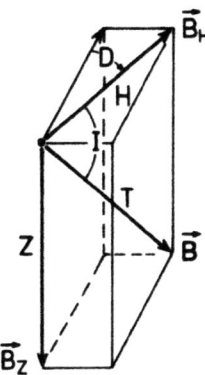

Illustration 4.1.1 – field elements

The so called **field elements of the magnetic field** can be generated from this vector. So, the declination, inclination, and the total intensity, etc.

The equations of Gauß and weber for the earth magnetic field are used till this day.

From the potential equation or also the theory the really appearing coefficients cannot be determined in the potential equation. This happens about a **measurement** of the real magnetic field of the earth.

Comment:
If one dissolves the brackets in the potential equation, merely products of sine or cosine functions appear in the equation of Gauß, so tesseral spherical harmonics or grids.
That is: Already Gauß and Weber based their considerations of the earth magnetic field on **spherical harmonics**!!!

4.2 – Measuring stations

The magnetic field of the earth is recorded worldwide by more than 200 measuring stations.

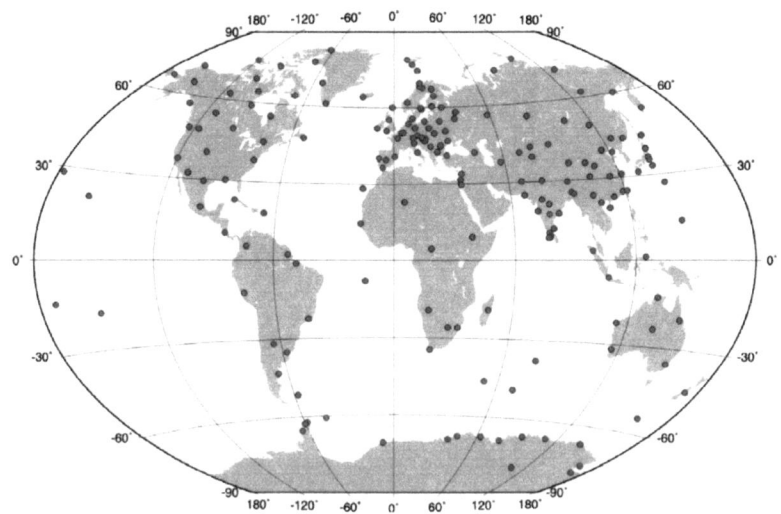

Illustration 4.2.1 – measuring stations

For some decades the field will also measure from satellites. The first satellite was Magsat which registered in 1980 during six months the intensity and the whole magnetic field.

Since 1999, the Danish satellite Ørstedt is located on an orbit and since July 2000, the German Champ satellite works.

All accumulated data about the earth's magnetic field are recorded by the IUGG and the ÌÅGA and evaluated. These values are used among others as a basis for creating the models **IGRF** and **WMM**.

These models are issued as flashcards for declination, total intensity, largest total changes, horizontal intensity, inclination, intensity of North, East intensity and vertical intensity.

4.3 – Total intensity – WMM 2005

The "world magnetic model" (WMM) is a product of the U.S. National Geospatial-Intelligence Agency (NGA).

You can relate the WMM 2005 about the „National Geophysical Data Center" (NGDC) now also via the Internet.

The NGDC and the British Geological Survey (BGS) made the WMM with the support of the NGA in the United States, together with the „Defence Geographic Imagery" and the „Intelligence Agency" (DGIA) from Great Britain.

The WMM is used as a standard model for the US Department of Defense, the UK Ministry of Defence, the North Atlantic Treaty Organization (NATO), and the World Hydrographic Office (WHO).

It is widely used in the civilian navigation. The model, the accompanying software and documentation are managed by the NGDC and the NGA. The model is made for every 5 years.

The card of the total intensity is important to this investigation, as a particular type of evaluation is possible.
So the total field total intensity at the earth's surface for the **World Magnetic Model 2005** can be seen in the following illustration 4.3.1.

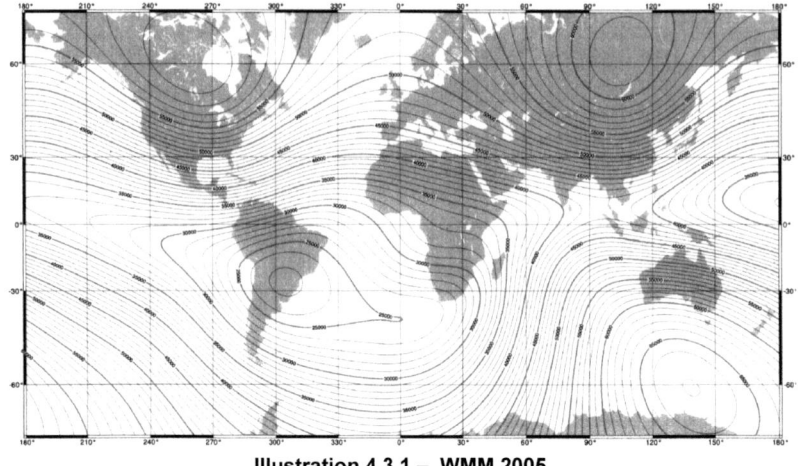

Illustration 4.3.1 – WMM 2005

The **four** extremes of the field here are striking (and not just two as a rod magnet or dipole model), where three Maxima and a minimum exist. In addition a saddle point appears (in the region of Indonesia)

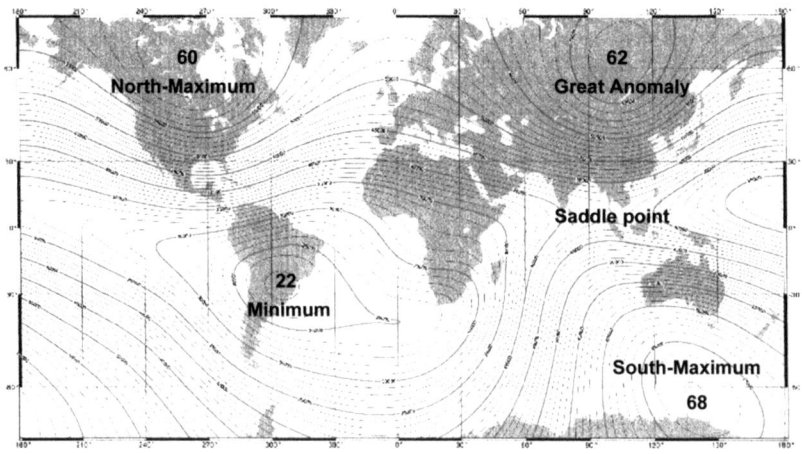

Illustration 4.3.2 – extreme values of the field

The names of the extreme values of the magnetic field is so commonplace in geophysics.
The formation of four poles is no longer explained by the dipole theory.

And when using the multi pole model:

Through the conversion of the terms in an infinite sequence, the whole process is a **mathematically approximate**, the solution only a **mathematical approximation**.
The question of **physical relevance** remains completely open! It is therefore right to doubt that nature here has used the same implementation.

4.4 – Temporary stability

The whole magnetic field changes its shape - in the long term. This slow change is called secular variation and results in the course of time that the magnetic field in its polarity reverses.
According to Geoforschungszentrum Potsdam, the last pole reversal is back about 750,000 years.

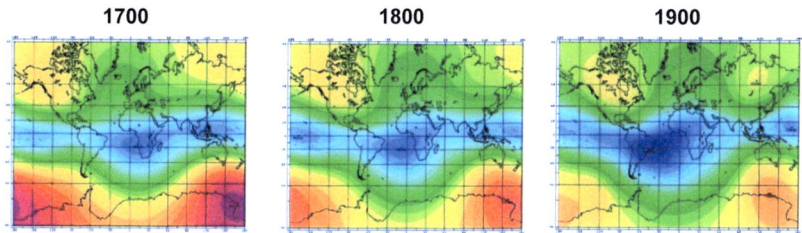

Illustration 4.4.1 – earth magnetic field

In the illustrations of the „World Data Center for Geomagnetism" in Kyoto, the situation for the last three hundred years can be seen. These images are based on geological findings and long-term back calculations of the earth's magnetic field.
Comparing the illustrations with the today's models from 2005, so virtually no change in the **structure** can be determined. Intensities change and also the extremes change their location, but also no major change in the **field structure** can be determined.

The field of total intensity is so temporary over time more or less constant.

It is therefore ideally suited for a general analysis of the earth's field.

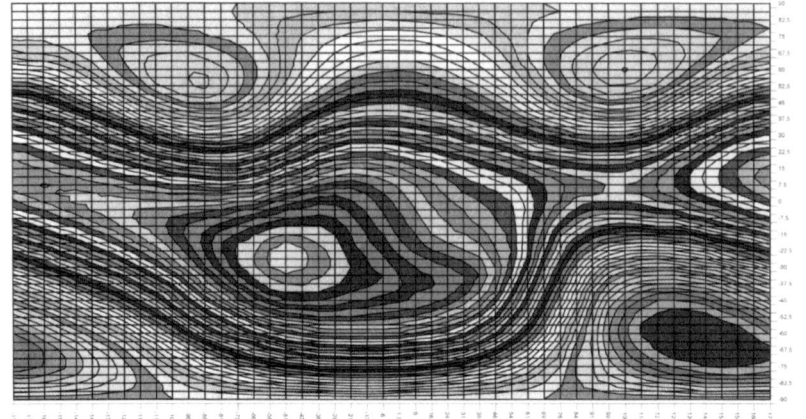

Illustration 4.4.2 – total intensity 1980

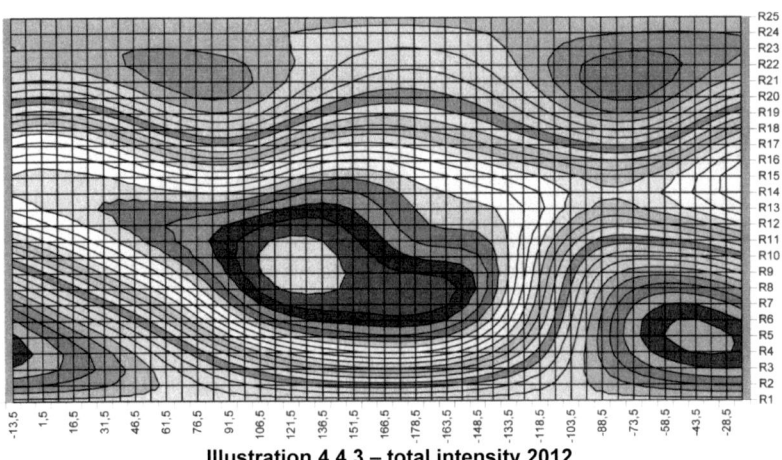

Illustration 4.4.3 – total intensity 2012

The comparison of the illustrations shows that there are no major differences between 1980 and 2012 in the structure of the field.
The magnetic field (total intensity) has remained sufficiently constant during the past 30 years.

4.5 – Fourier-analysis of the earth magnetic field

The approach here shall find an oscillation base for the earth's magnetic field.
It exists a mathematical method, namely the Fourier analysis. With it a given periodic function can be decomposed as a sum of sine and cosine functions.

Illustration 4.5.1 – total intensity

The illustration 4.5.1 is converted into a **table** of total intensities. And so, the possibility of an evaluation is given.
Also a 3D representation of the earth's magnetic field at the earth's surface is possible.

Here, the evaluation is made via a **two-dimensional Fourier analysis.**
A Fourier analysis provides a decomposition of a given function as a sum of sine and cosine functions. A two-dimensional Fourier analysis receives a sum of **spherical harmonics**.

And with the earth's magnetic field actually can be made a decomposition into spherical harmonics.

4.5.1 – Fourier-analysis

The total field is decomposed into single - in this case horizontal – **cuts**. With each cut, a one-dimensional Fourier analysis can be done.

Each section runs along a **parallel**. The cuts are created by + 90 North to -90 South at a distance of 7.5 degrees. The results are 25 sections. (7.5 Degrees = 800 km)

Considering a one-dimensional (numeric) Fourier analysis to every cut with the variable λ and the step of 7,5 degrees, i.e. with 48 points per cut. There are total 1106 Points for the analysis of the earth's field.

A **numeric** (one-dimensional) harmonic procedure serves as a basis, or an analysis tool, as described in the book „Mathematics for engineers" by Brauch/Dreyer/Haacke and as **Algorithm of Goertzel** (and Reinsch) is known.

Then, the 25 slices from the Fourier analysis generally can be represented:

$$Y_0 = \sum_{i=0}^{n}(a_{i0} \cdot \cos i\lambda + b_{i0} \cdot \sin i\lambda) \quad \text{bis} \quad Y_m = \sum_{i=0}^{n}(a_{im} \cdot \cos i\lambda + b_{im} \cdot \sin i\lambda)$$

The result is a system of equations with **m+1** equations with each **n+1** elements.

The generated coefficient matrix with the **A_m** and **B_m** can be performed as **one more** Fourier analysis, with the variable φ. Every point on the earth with the coordinates λ, φ qualifies as:

$$Y = \sum_{j=0}^{m}(a_{j0} \cdot \cos j\varphi + b_{j0} \cdot \sin j\varphi) + \sum_{i=1}^{n}\sum_{j=0}^{m}(a_{ji} \cdot \cos j\varphi + b_{ji} \cdot \sin j\varphi) \cdot (\cos i\lambda + \sin i\lambda)$$

$$Y_0 = a_{00} + a_{10} \cdot \cos\lambda + b_{10} \cdot \sin\lambda + a_{20} \cdot \cos 2\lambda + b_{20} \cdot \sin 2\lambda + \ldots + a_{n0} \cdot \cos n\lambda + b_{n0} \cdot \sin n\lambda$$

$$Y_k = a_{0k} + a_{1k} \cdot \cos\lambda + b_{1k} \cdot \sin\lambda + a_{2k} \cdot \cos 2\lambda + b_{2k} \cdot \sin 2\lambda + \ldots + a_{nk} \cdot \cos n\lambda + b_{nk} \cdot \sin n\lambda$$

$$Y_m = a_{0m} + a_{1m} \cdot \cos\lambda + b_{1m} \cdot \sin\lambda + a_{2m} \cdot \cos 2\lambda + b_{2m} \cdot \sin 2\lambda + \ldots + a_{nm} \cdot \cos n\lambda + b_{nm} \cdot \sin n\lambda$$

A second numerical harmonic analysis is performed on the column coefficients:

$$Y = \sum_{j=0}^{m}(a_{j0} \cdot \cos j\varphi + b_{j0} \cdot \sin j\varphi)$$

$$+ \sum_{j=0}^{m}(a_{j1} \cdot \cos j\varphi + b_{j1} \cdot \sin j\varphi) \cdot \cos\lambda$$

$$+ \sum_{j=0}^{m}(a_{j1} \cdot \cos j\varphi + b_{j1} \cdot \sin j\varphi) \cdot \sin\lambda$$

$$+ \sum_{j=0}^{m}(a_{j2} \cdot \cos j\varphi + b_{j2} \cdot \sin j\varphi) \cdot \cos 2\lambda$$

$$+ \sum_{j=0}^{m}(a_{j2} \cdot \cos j\varphi + b_{j2} \cdot \sin j\varphi) \cdot \sin 2\lambda$$

+

+

$$+ \sum_{j=0}^{m}(a_{jn} \cdot \cos j\varphi + b_{jn} \cdot \sin j\varphi) \cdot \cos n\lambda$$

$$+ \sum_{j=0}^{m}(a_{jn} \cdot \cos j\varphi + b_{jn} \cdot \sin j\varphi) \cdot \sin n\lambda$$

The first term for Y (the zonal, sectorial part) can be still complemented so:

$$\sum_{j=0}^{m}(a_{j0} \cdot \cos j\varphi + b_{j0} \cdot \sin j\varphi)$$
$$= \sum_{j=0}^{m}(a_{j0} \cdot \cos j\varphi + b_{j0} \cdot \sin j\varphi) \cdot \cos 0 + \sum_{j=0}^{m}(a_{j0} \cdot \cos j\varphi + b_{j0} \cdot \sin j\varphi) \cdot \sin 0$$

Due to the quantitative analysis on the one hand and the mathematical methods, on the other hand, the following total equation for the magnetic flux density at the earth's surface can be created:

4.5.1.1 - Equation:

$$B = \sum_{i=0}^{n} \sum_{j=0}^{m} \left(a_{ji} \cdot \cos j\varphi + b_{ji} \cdot \sin j\varphi\right) \cdot \left(\cos i\lambda + \sin i\lambda\right)$$

If one solves the brackets, just **tesseral spherical harmonics**, so **grids**, appear.

4.5.1.2 - Theorem:
> The magnetic field of the earth (on the earth surace) can be completely described by a sum of grids.

At the same time, this function represents a solution of the angle part of Laplace's equation. In the consequence is then:

4.5.1.3 - Theorem:
> Magnetic field of the Earth (on the surface)
> = two-dimensional oscillation structure

Here arises the question whether there is a relation between the two-dimensional magnetic oscillation model and the earth oscillation structure?

Because of the general approach of the Fourier analysis and of the equation 4.5.1.1 the following general statement can be made:

4.5.1.4 - Theorem:
> Each oscillation structure on or around a sphere can be completely described by a sum of tesseral spherical harmonics.

Considering the statements and equations in chapter 2.11.1 the following can be said:

4.5.1.5 - Theorem:
> Each oscillation phenomenon around a sphere can be represented as a solution of the angular part of the Laplace equation.

4.5.2 – Quantitative Fourier-analysis

The quantitative **numeric** Fourier analysis of the earth's magnetic field (IGRF 1984) yields the following result for the magnetic flux density at the earth's surface:

4.5.2.1 - Equation:

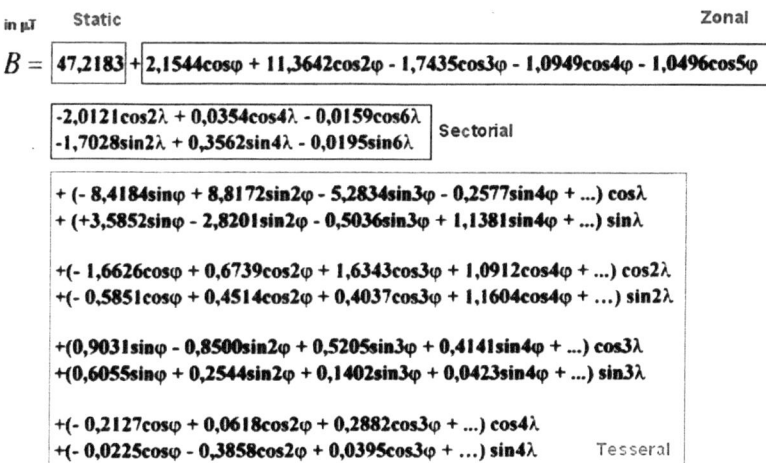

The evaluation provides all three kinds of spherical harmonics, so zonal, sectoral and tesseral forms, as well as a static part.
The zonal and sectoral part of the spherical harmonics can be grasped to one grid, called the **grid ZS**. This behaves then like a tesseral spherical harmonic, so as a grid.

4.5.2.2 - Equation: Grid ZS = Zonal + Sectorial

$$B_{ZS} = B_z + B_S$$

Comment:
In the equation 4.5.2.1 exist 17 intensity values that lie between one and eleven micro-Tesla. All other values are in the range of **nano-Tesla**!!!

4.6 – Further evaluations

The individual parts of the Fourier analysis allow a graphical representation of the total situation.

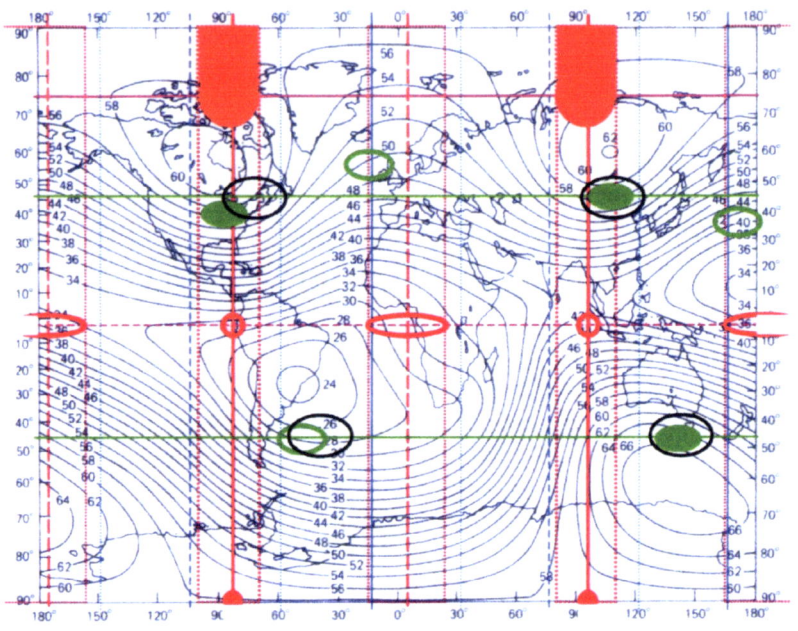

Illustration 4.6.1 – earth magnetic field

The map of the total intensity registered all extremes, magnetic structures and source points arising from the analysis.

Blue – three-axle ellipsoid
Red – Zonal, Sectorial (Grid ZS)
Green – Tesseral
Black – Hyugens source points

The illustration 4.6.1 entails relationships that are represented in detail on the following pages.

4.6.1 – Three-axle ellipsoid

With the help of satellite geodesy by C.A. Lundquist and G. Veis in 1966 the following parameters have been identified to represent the earth as a genuine three-axle ellipsoid:

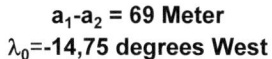

$a_1 - a_2$ = 69 Meter
λ_0 = -14,75 degrees West

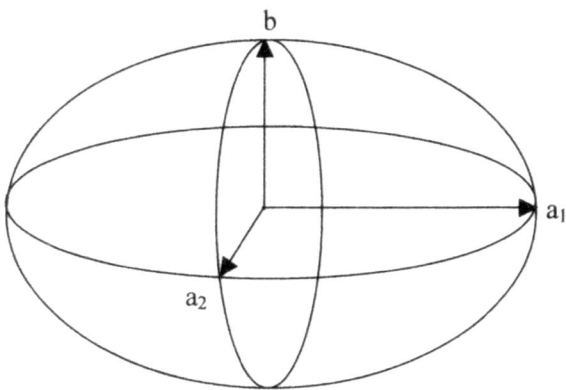

Illustration 4.6.2 – three-axle ellipsoid

The blue ellipsoid grid in illustration 4.6.1 is based on the values of Lundquist and Veis and is moved to the red magnetic system about 1.25 degrees.
So **seen globally**, a good match exists.

An analysis of the geographic locations of all occurring magnetic extremes results in a functional relationship for their longitude. The derivation of the following equation can be found in the book "Lattice structures of the earth magnetic field" Chapter 4.4 and 9.5.

4.6.1.1 - Equation: $\qquad \lambda_E = 3{,}75° \cdot m - 13{,}5°$

m is element of the integers (...-2,-1,0,1,2,...)

4.6.1.2 - Theorem: The earth magnetic field stand in relation to the figure of the earth.

4.6.2 – Grid ZS

The red magnetic system in the illustration 4.6.1 represents the **grid ZS**, so the zonal-sectoral part.

Creating a magnetic back at the north pole, while there is only a maximum point at the south pole.

The maximum zone = **magnetic main meridian** (thick red) is clearly visible, with lambda = -83.5 degrees west and lambda = 96.5 degrees east.

The minimum zones (red dashed line) are at lambda = 5.25 degrees east and lambda = -174.25 degrees west.

There are two minimal zones and two saddle points in the equatorial plane.
.

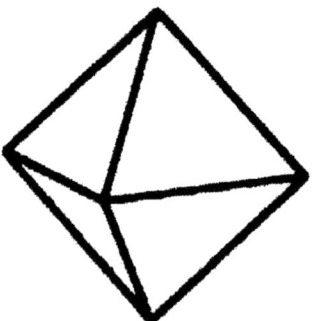

Illustration 4.6.3 – octahedron

4.6.2.1 - Theorem: All extreme value areas of the grid ZS lie on the corners of an octahedron.

4.6.3 – Tesseral field

The green system in illustration 4.6.1 provides the tesseral part. All extreme values lie approximately ±45 degrees latitude.

The green points represent the maximum and minimum points of the pure (tesseral) grid part of the earth's magnetic field.

Solid green = Maximum
Green rimmed = Minimum

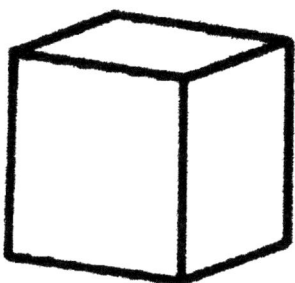

Illustration 4.6.4 – cube

In the northern hemisphere all extremes lie approximately on a **square**. Through the 45 degrees latitude a twisted **parallelepiped** (cube) is created in the earth, in terms of coordinate.
The extremal zones in the southern hemisphere are about 35-40 degrees **moved** against the northern extremal zones.

4.6.3.1 - Theorem: The extreme values of the tesseral field are located on the corners of a twisted cube. (parallelepiped)

4.6.4 – Huygens source points

The **black** rimmed ellipses represent the source points of the total field.

The basic field, the grid model and the Huygens principle provided that these four poles represent the **theoretical source points** from which the **whole outer magnetic field at the earth's surface** can be created

There is alignment of the source areas with the four main extrema of the tesseral field.

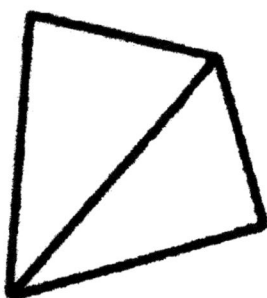

Illustration 4.6.5 – tetrahedron

The source points are located on the corners of a **tetrahedron**. The sources in the southern hemisphere are shifted by 45 degrees.

4.6.4.1 - Theorem: The Huygens source points are located on the corners of a distorted tetrahedron.

4.6.5 – Summary

All extreme value areas of the grid ZS are on the corners of an **octahedron**.

The extreme value of the tesseral field are located on the corners of a twisted **cube** or a **spar**.

Comment:
A parallelepiped (Synonyms: spar, Parallelotope) is a geometric body bounded by six parallelograms. The name spar comes from the calcite, whose cristals have the shape of a paralelepiped.

The Huygens source points of the field are located on the corners of a distorted **tetrahedron**.

The earth magnetic field stand in relation to the figure of the earth. An analysis of all magnetic **extremes** results in a functional relationship for their longitude:

$$\lambda_E = 3{,}75° \cdot m - 13{,}5° \qquad 3{,}75 \Leftrightarrow 96 \text{ Division}$$

With the 96 Division, a sufficient differentiation exists to contain **all** occurring angle for polyhedra or the Platonic solids.

4.6.5.1 - Theorem: All Platonic solids are available as oscillation figures of the earth oscillation structure.

4.7 – Huygens source points of the earth field

As listed in Chapter 2.2.6, there are two ways to see a grid. The grid can be described in its own level of development.
This happens for example in the Fourier analysis as seen. Or one can describe the grid on the plane of basic oscillations. This will happen in the following.

4.7.1 - Ideal source point structure

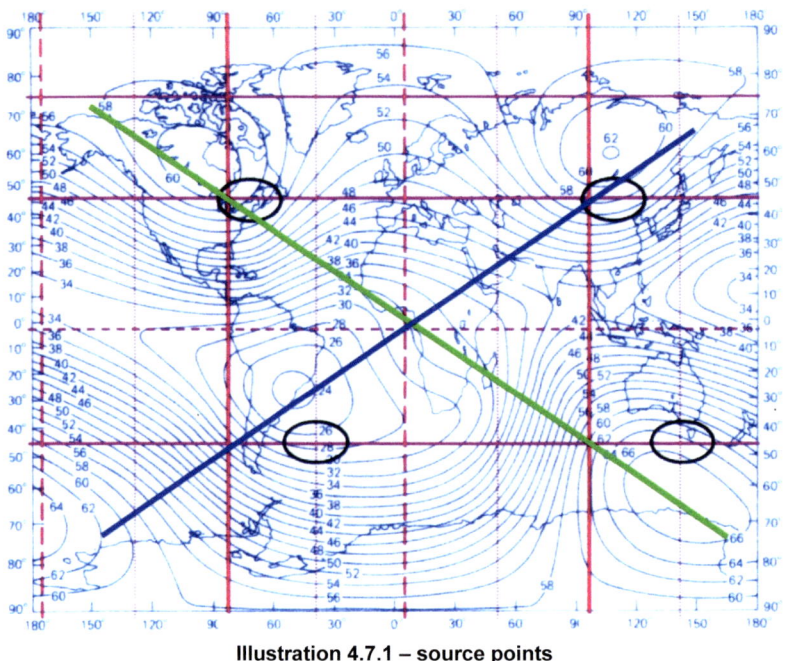

Illustration 4.7.1 – source points

The picture shows the determined Huygens source areas of the total field (black) with the magnetic main meridian (thick red vertical)

The blue and green lines represent the theoretical (mathematical) connection between ideal source points.

It is expected that the location of the extremes of the total intensity on these lines are located and the extremes lay in the vicinity of the source points - where the source points are, also the largest intensities should occur.

It can also be seen that there is a variance (disturbance) from the ideal configuration in the southern hemisphere !!!

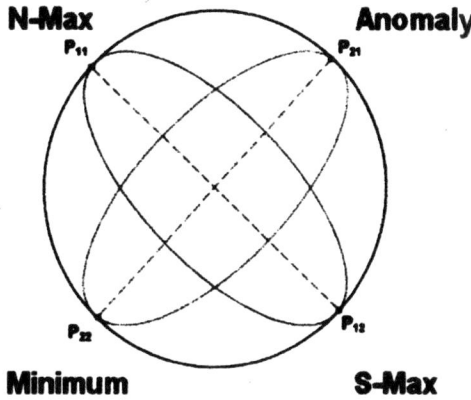

Illustration 4.7.2 – extrema -structure

On the N-Max - S-Max axis an oscillation builds up, which behaves like a **straight** oscillation - face two Maxima.
On the minimum anomaly-axis an oscillation builds up, which behaves like an **odd** oscillation - face a maximum and a minimum.
The circle that connects all source points stands vertically on two levels and is identical with the magnetic main meridian.

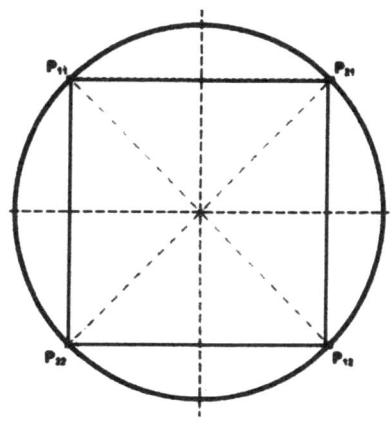

All sources in the plane of the main meridian on the corners of a **square** build an ideal (undisturbed) source point arrangement.

Illustration 4.7.3 – ideal arrangement

4.7.2 - Real source point structure

In the northern hemisphere, the pole or sources are located on the magnetic main meridian, in good agreement with the undisturbed source point arrangement. In the southern hemisphere, however, exists a deviation from 45 degrees to the east.

There are two possibilities of the source point arrangement:

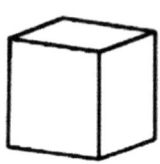

The source points in the northern hemisphere are opposite diagonally on the cube. You can choose the source points of the southern hemisphere now:
either at the bottom of the cube vertical under the northern points or at the bottom of the cube with 90 degrees offset.
In both cases, the lower part of the cube is twisted then to ±45 Grad degrees to get the real positions.

The source points in the northern hemisphere are opposite above on the tetrahedron.
The source points of the southern hemisphere face down on the tetrahedron.
The lower part of the tetrahedron, is twisted to 45 degrees westward, to get the real positions.

Illustration 4.7.4 – source points

4.7.2.1 - Theorem: Spatial order structures for the real (Huygens) source points are polyhedra.

5.0 – Generating and generated elements

Flows of liquid magma in the earth's core are viewed as the physical cause of the earth magnetic field. The outer core lies at a depth between approximately 2,900 Km and 5,100 Km, or a center distance of 1.271 Km to 3.471 Km. There rotates a liquid ball shaped mass of an iron-nickel mixture around the core. This mass can be understood as electric moved charges which generate magnetic pulsating fields as a result. The wobbling movement of the earth around the sun with the inclined axis creates a "stirring effect" around an elliptical center. This creates the pulsations of the field.

The earth magnetic field producing elements are magmatic flows of approximately 2900 km depth.

These flows are too slow and too strong to be affected by short-term geological or solar events. The earth magnetic field, the magnetic oscillation structure, and thus the magnetic grid system have therefore a certain inertia that impact contrary to all external influences.

Change of the earth magnetic field in intensity and structure can be achieved only through change of magmatic flows in their direction or their flow velocity or density. Possible factors are the Coriolis force, chemical and thermal convection and the repercussion of the generated magnetic field on the earth's core.

The existence of four poles with regard to the total intensity indicates two currents or flow systems. The disturbance of the earth magnetic field in the southern hemisphere shows that there (at least) two currents and flow systems exists which are not properly balanced to each other and are also not synchronized or tightly coupled.

If there exist two magnetic field generating flow systems so there are four poles. This results in three possible structures of oscillation:

a) build by two straight oscillations - four maxima

b build by two odd oscillations - two maxima and two minima

c) build by a straight and an odd oscillations - three maxima and a minimum.

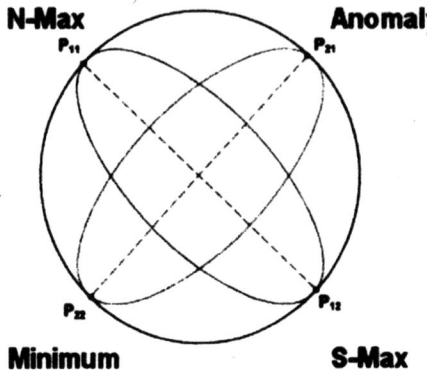

Illustration 5.0.1 – oscillation planes

The picture shows the four poles and the associated oscillation planes.
The circle through all four poles is the main meridian of the whole system.

Due to the laws of physics, this basic oscillations should begin on the **core ball**, on which the magnetic field produces currents move.

5.1 – Core balls

The oscillation structure provides an approach for the core ball. Assumption is the layer equation 2.5.2.2.

For **n = 1** exist a simple relation between the radius of the (producing) ball and the center distance of the generated layers. If one sets radius = 1 (and thus standardizing it) one get:

n	k m	1	2	3	4	5	6	7	8
1	1	1	3	5	7	9	11	13	15

For k = 1 always the radius of the producing ball appears. Two cases are of interest here:

5.1.1 - CASE 1

The first case is k=2. **The proportionality factor is three and one third**. An oscillation structure compatible to the basic hull is created by a core sphere which has a radius that is one third of the basic hull radius.

Basic hull = L_o = 6.355,76 Km   one third = 2.118,59 Km

This corresponds to a depth of 4.252,41 Km. The **outer** core is in a depth between approximately 2,900 km and 5,100 miles. The core sphere is so located in the outer core, which is exactly in the zone in which the magnetic field generating magmatic flows are.

5.1.1.1 - Theorem: **The magnetic field generated by the magmatic flows in the outer core has an oscillation structure compatible to the earth oscillation structure.**

The basic frequency of the core ball with 35,37 Hz is three times bigger than the earth basic frequency.

5.1.2 - CASE 2

The second case is k=3. **The proportionality factor is five and a fifths**. An oscillation structure compatible to the basic hull is created by a core sphere which has a radius that is a fifths of the basic hull radius.

Basic hull = L_o = 6.355,76 Km   one fifths = 1.271,15 Km

This corresponds to a depth of 5.099,85 km. The **inner** core is in a depth between 5,100 km and the centre at 6.371 km below the earth surface. This core sphere with fifths radius is identical to the inner core and can be considered as inner core ball.

5.1.2.1 - Theorem: **The oscillation structure generated by the inner core is compatible to the earth oscillation structure.**

The basic frequency of the inner core ball is five times larger than the basic earth frequency and is 58,96 Hz.

The three oscillation structures that depend on the basic hull, the outer core and the inner core are in **harmonic** proportions and are identical in parts of their grid structure, they have the **same oscillation structure**, which is **equivalent** to the earth oscillation structure.

5.1.3 - Theorem: inner core : D"-Layer : basic hull = 1:3:5

Comment:
The D"-layer forms the lowest part of the lower mantle and thus represents the transition zone between the mantle and the core. The thickness of this layer is between 200 and 300 km.

Simplified can be written:

5.1.4 - Theorem: inner core : outer core : earth radius ≈ 1:3:5

Comment:
The ratio 3:5 or 5:3 is located close to the golden ratio. You could therefore say: **basic hull and D"-layer are roughly in the ratio of the golden section.**

Simplified can be written:

Earth radius and outer core are roughly in the ratio of the golden section to each other.

5.2 – Creation of geological layers

The now widely accepted model for the formation of the moon says that a celestial body named Theia, with the size of Mars, nearly streaking with the proto-earth collided, about 4.5 billion years ago.
Theia itself was completely destroyed in this collision. The fragments from this impact gathered into an orbit around the earth.
Most of the impactor combined with the proto-earth to the Earth. After actual simulations the Moon was formed at a distance of approximately three to five earth radii, at an altitude between 20,000 and 30,000 Km.
The wreckage of the collision formed immediately (i.e. in less than 100 years) the proto-moon which gathered quickly all remaining debris and condensed, after nearly 10,000 years, to the moon with roughly today's mass. It circled the earth at that time at a distance of only about 60,000 km (double planet). That must have led to extreme tidal forces which deformed the egg-shaped earth and the moon.
The tidal forces were about 200 times stronger than it is today. Because the entire planet was an magmatic mass, and so was much more movable than today. Mass parts could be lifted by the tide lifting between 1 and 2 km.

Sole effects of rotation and gravity would occur in the earth to a continuous mass distribution, but not to a layer formation.

According to Chapter 3.5 is:
layer structure of the earth ⇔ earth oscillation structure

According to chapter 4.6.1 is: the earth magnetic field is in relation to the figure of the earth regarding a three-axle ellipsoid.

The three oscillation structures which caused by the basic hull, the outer core, and inner core, are identical with the earth oscillation structure according to Chapter 5.1.

The rotation period of the earth was at that time about 4 hours and was about 6 times faster than it is today. This had direct effect on the earth magnetic field, which was also about 6 times stronger than it is today.

Since the matter of the earth has **para/dia/ferro** magnetic properties and therefore magnetic fields can exert also **force effect** on the relevant matter, the formation of the geological layers can be explained if one assumes:

earth magnetic field ⇔ earth oscillation structure

The magnetic oscillation structure at the beginning of the earth building result in the forming of the geological layers served as crystallization basis for the liquid, magmatic matter.

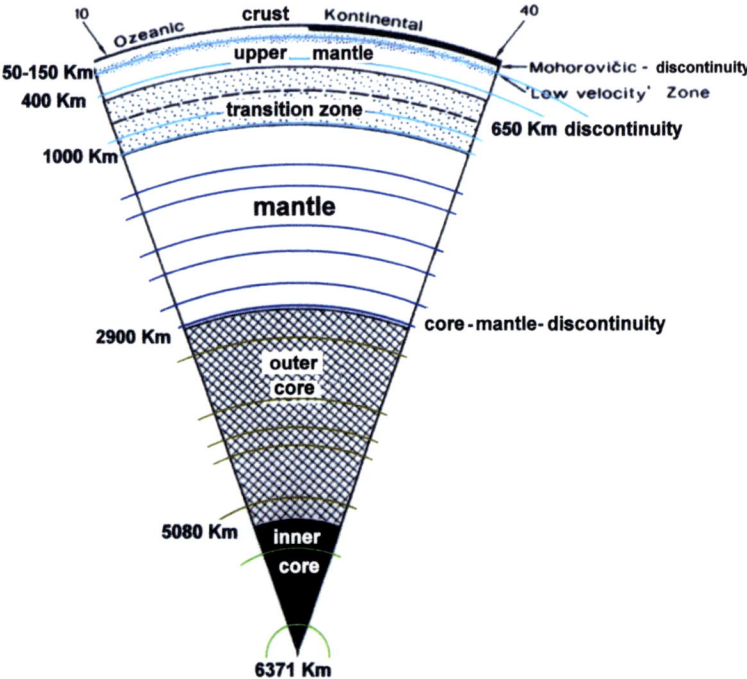

Illustration 5.2.1 – layers

5.2.1 - Theorem: **Layers**
= Clusters of maximum oscillation states
⇒ energy input

The formation of the geological layers can be explained as a resonance phenomenon.
The surrounding matter can be in response to the layers (and their frequencies) or not.

Where the surrounding matter was in response to the frequencies of the magnetic layers, there ensued energy input and force input and carried out a separation of the phases of matter.
Where the matter was not in response to the frequencies of the magnetic layers, was a crystallization in particular crystal forms.

5.2.2 - Implication: **The earth magnetic field and the magnetic oscillation structure exist since the formation of the core, and at the latest on the forming of the geological layers.**

According to chapter 4.7.2 applies: all polyhedra are possible oscillation figures.

That may have served as the basis of the crystal formation of magmatic matter by forming the geological layers.
Therefore, the polyhedron systems would be geological manifestation or crystallization of the oscillation structure at that time.

5.2.3 - Theorem: **Polyhedron systems are geological crystallization of magnetic oscillation states that prevailed by the forming of the core and the geological layers.**

5.3 – The electric field of the earth

According to chapter 3.6 applies:
layers of the atmosphere ⇔ earth oscillation structure

The ozone, d-, e- and f-region form the electrically more conductive layers of the atmosphere. This can be explained if one assumes:

earth magnetic field ⇔ earth oscillation structure

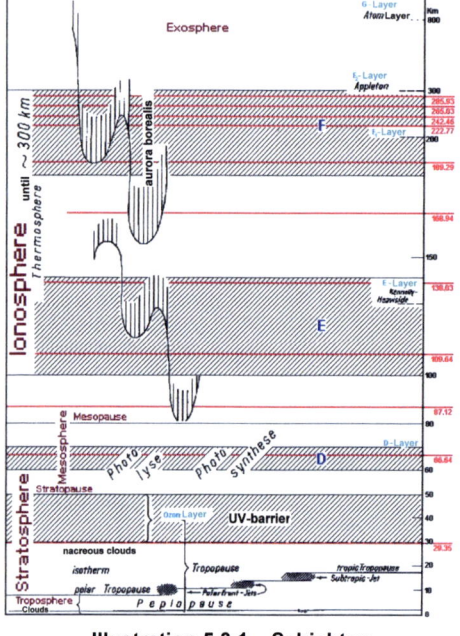

**Layers
= Clusters of maximum oscillation states
⇒ energy input**

In the layers, the following phenomena can occur:

1) electric Induction
2) Lorentz force

Consequence:
Increasing ionisation is generated due to the involved matter and this leads to the forming of electrical equipotential layers.

Illustration 5.3.1 – Schichten

The formation of the atmospheric layers can be interpreted as a resonance phenomenon through the increased ionisation of the involved matter with the magnetic layers.

5.3.1 - Theorem: The magnetic layers form the basis for the conductivity of the ozone, d-, e-, and f-region.

5.3.2 - Implication: The magnetic oscillation structure is the motor of the electric earth field.

Therefore, the equipotential layers of the electric earth field are constantly recharged by the earth magnetic field. And only this correlation, stabilizes the electric earth field.

Due to the relations between atmosphere, climate and weather, the earth magnetic field has two action moments, the climate regarding:

1) through the formation of electric more conductive layers and the electric field

2) through direct exposure (Lorentz force) on water and air mass transport

These very small forces may be local, but because they are effective anywhere on earth and would have some influence on the global air / clouds / water flows. The relationship of earth magnetic field and climate is not considered in any of the existing climate and weather models.

5.3.3 - Implication: The magnetic oscillation structure is the motor of the earth climate.

It make sense to ask what influence probably would have a **change** of the earth magnetic field on the atmosphere and their processes.
There is a certain probability that the climate change of the past years partly induced by the magnetic field also changing.

5.3.4 - Implication:

change of earth magnetic field $\Rightarrow$ change of earth climate

The consequence is that the human being and its emissions have an accelerating effect at the climate change, but are not the cause of climate change.

5.4 – One oscillation structure

I) After chapter 4.4.2 is:
magnetic field on the earth surface = a two-dimensional oscillation structure

II) After Chapter 3.5 is:
layer structure of the earth ⇔ earth oscillation structure

III) According to Chapter 5.0 the forming of geological layers can be explained by a magnetic oscillation structure.

IV) After chapter 3.6 is:
layers of the atmosphere ⇔ earth oscillation structure

V) According to Chapter 5.1 the forming of the electric earth field can be explained by a magnetic oscillation structure.

Consequence:

    earth magnetic oscillation structure ⇔ earth oscillation structure

5.4.1 - Theorem: **The earth with their inner structure, the atmosphere, their magnetic field and the electric field can be described by a single earth oscillation structure.**

The earth magnetic field takes a key role, because it is the medium which manifests the oscillation structure in material and energy in the earth system.
Since the (magnetic) oscillation structure has a significant part in the physical implementation, the underlying model as well as the Fourier analysis based on oscillations and the previously observed physical events are consistent with the theoretical model, is as follows expressed the **core hypothesis** for the physical implementation of the model here:

5.4.2 - Theorem: **Standing earth magnetic waves have physical reality.**

For the determination of the existence or non existence of these waves, still a measurement procedure is specified in this book. This then represents the experimentum crucis of this core hypothesis 5.4.2.

5.5 – Sub structure

The earth magnetic field producing elements are magmatic flows in the outer core. Due to continuous repetition of pole reversal over the last billion years and also ongoing reconstruction of the field, suggests that within the magnetic field generating currents are also cyclical processes. (the period is probably close to a multiple of the precession)

Physical moments such as the Coriolis force, chemical and thermal convection and the repercussion of the generated magnetic field on the core seem to be the basic motor, which holds the magmatic currents in the liquid outer core, leading to the movement of the geodynamo.

According to chapter 4.8, two flow systems are available which are not properly balanced to each other and are not tightly coupled or synchronized. Since the geodynamo again and again is accelerated, the consequence is:

5.5.1 - Theorem: Every time after a polarity reversal, the magnetic oscillation structure builds up again.

5.5.2 - Implication: The magnetic oscillation structure is a sub structure of the whole earth magnetic field.

5.5.3 - Implication: The equations from Gauß and Weber are not enough to describe the fully earth magnetic field.

The equation of Gauß and Weber describes the vector nature of the field, but not the character of stratification.

Part 4 – Demonstrability

With the construction of the oscillation structure in part 1 a mathematical and physical model is available that allows to explain the structures of the earth on a basis of oscillations.

In part 2 chapter 3, the model is applied to the so called classical planetary systems of the earth – that means also the geological layers as well as the layers of the atmosphere.

In part 3, chapters 4 and 5 the application of the model was applied to the earth magnetic field and the electric field of the earth.

The following chapters are about the provability or falsifiability of the present model.

6 – Measuring of magnetic waves

6.1 – Classic Hall sensor

To measure magnetic fields, you need a physical effect, that responds to the magnetic flux density B. The Hall effect provides such a phenomenon.

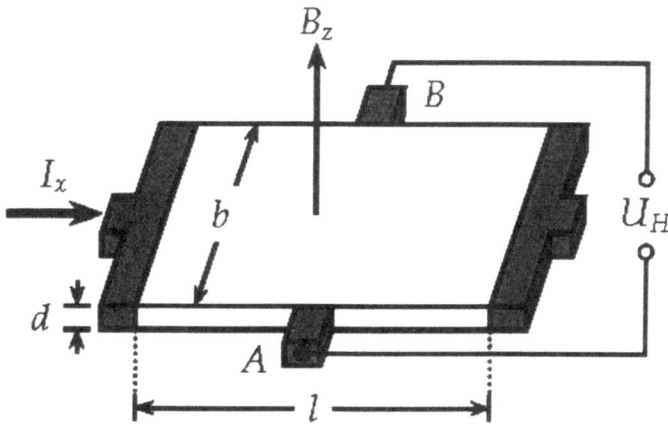

Illustration 6.1 – Hall sensor

As you can see in the illustration 6.1, a very thin metal plate is crossed by a uniformly distributed current I_x. No voltage is between the two points A and B, which are equidistant from the electricity supply and are connected to a galvanometer.

The plate is interspersed by a magnetic field B, the Lorentz force F appeals on the moving charge carriers of the current I. This force repeals the uniform charge distribution in the plate and leads to a potential difference U_H between the two points A and B. A current flows through a galvanometer connected to these points.

The charge distribution induced by the Lorentz force generates an electrical field with the field strength E, which counteracts the distraction of the charge carriers. A state appears, where the Lorentz force of F_L and the force caused by the electric field F_E has the same amount.

The general equation for the Hall Voltage can be derived from these boundary conditions:

6.1.1 - Equation: $$U_H = R_H \cdot \frac{I \cdot B}{d}$$

Where the Hall coefficient **R_H**, as well as the thickness of the metal plate **d** as specific constants of a Hall sensor are to be considered, summarized in a constant **k**. To be written as:

6.1.2 - Equation: $$U_H = k \cdot I \cdot B$$

The Hall voltage is practically only depending by the power supply I of the sensor and the flux density B of the applied magnetic field. When the Hall effect sensor operates with a DC voltage then I is constant.

In this configuration, the Hall sensor is typically used as a flux density meter.

6.2 - New functionality

Strictly speaking the current I and the flux density B can including time varying sizes in the equation of the Hall voltage. It is limited (first) to sine forms, resonance effects caused by the used frequencies, can be used to find application as a measuring device.

The Hall effect is usually operated with DC voltage, although the equation allows other types of voltages. So also sinuidale forms. Taking for the current I and the magnetic field B sinus forms you get the following general equation for the Hall voltage:

6.2.1 - Equation: $$U_H = k \cdot \hat{I} \cdot \hat{B} \cdot \sin(I) \cdot \sin(B)$$

The Hall Voltage is the product of two sine waves here. This can be used also electro technically implemented and so that magnetic waves can be recorded.

The novelty is to adjust the Hall sensor on a defined working point and then to provide an AC voltage - like a transistor. The result is the **ejPi measurement method**.
The Ebbers-Jähn-Piontzik-measuring procedure, in the following short **ejPi measuring** called or after its producers called also **ejPi-method**, allows the frequency measurement of magnetic waves.

By the ejPi-method the Hall sensor is supplied with a DC voltage. (**half** operating voltage) A variable AC voltage is added to it.

If you now bring the sensor in a sinusoidal magnetic field, the generated Hall voltage UH is a **multiplication** of two existing waves (current and magnetic field).
The resulting wave is usually a beat form, as shown in the following illustration 6.2.

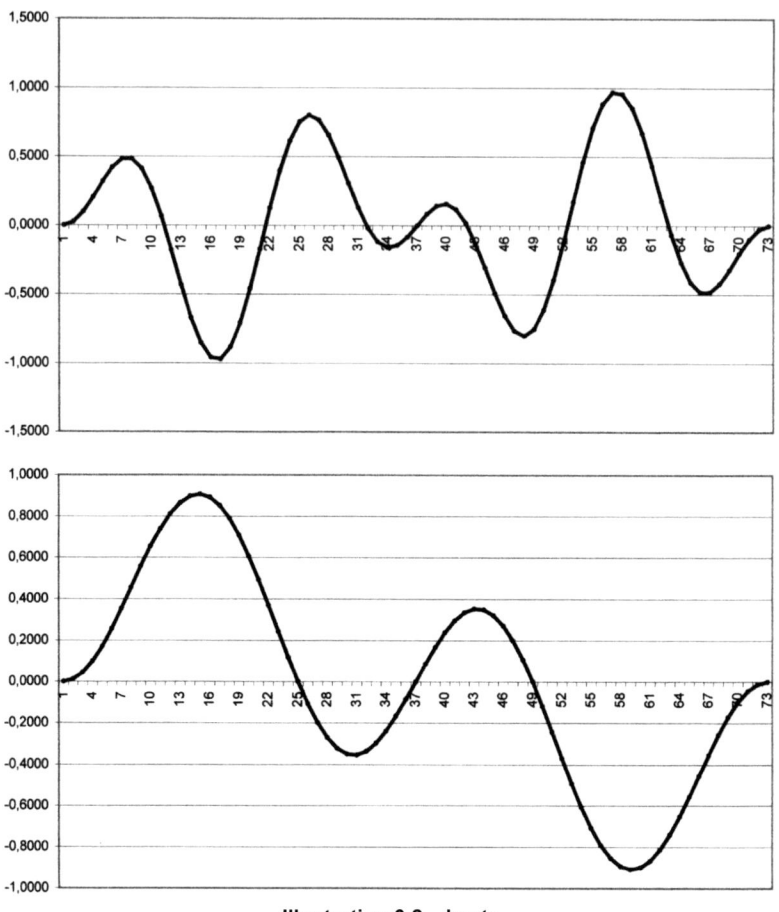

Illustration 6.2 – beats

A kind of resonance effect occurs at same frequency of the supplied current and the applied magnetic field - the result is a clean sine square wave without negative voltage components, as shown in Figure 6.3.

6.2.1 - Equation: $\quad U_H = k \cdot \hat{I} \cdot \hat{B} \cdot \sin^2(\omega \cdot t)$

The addition theorem applies to trigonometric expressions:

$$\sin^2 \alpha = \frac{1}{2} - \frac{1}{2} \cos(2\alpha)$$

The result is a clean cosine wave.

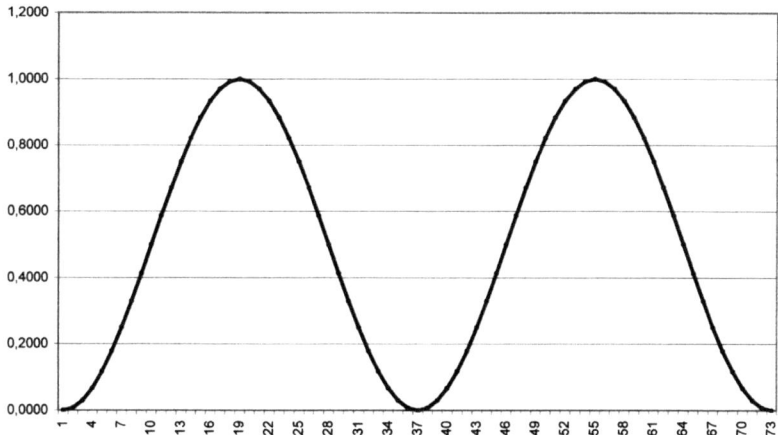

Illustration 6.3 – sine-square-function

This resonance reaction can be evaluated visually through a measuring instrument (oscilloscope) or switching on other electronic and digital components, and allows the measuring of the resonant frequencies.

It is in any case a Hall effect sensor required with the classic four connections and generating the analogue Hall effect. Modern integrated Hall sensors with three connections are disabled here.

The **ejPi-method** allows the frequency measurement of magnetic waves. This can be metrologically used and in the fields of testing, tracking, and geophysics. The procedure is registered under the number 102012011759 as patent at the German Patent and Trademark Office.

6.3 – Circuit to the measuring procedure

The circuit for the ejPi measurement method essentially consists of three parts:

a) the Hall sensor
   For the ejPi measuring principle each Hall sensor can be used, which has four connections and generates the analogue Hall effect.

b) generation of variable supply voltage for the sensor
   The Hall sensor is supplied with a DC voltage on which an adjustable or variable AC voltage is added.

c) buffering and amplification of the sensor output signal
   The sensor output signal is buffered by a differential amplifier.

The circuit is designed for static as well as dynamic supply voltages (with static working voltage).

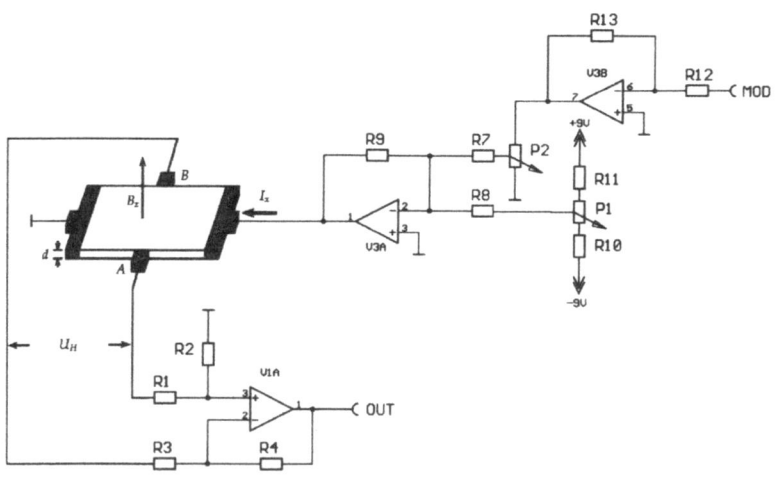

Illustration 6.4 – circuit for the ejPi-measuring procedure

6.4 – Experimentum crucis

The experimentum crucis for the magnetic waves looks as follows:

basis of the experiment is the ejPi measurement method

The Hall sensor is powered by an AC component. This AC voltage can be **adjusted over the whole frequency spectrum**.

If geomagnetic waves exist, then resonances at certain frequencies (like the earth frequency or its harmonics determined in Chapter 3) must occur.
With this experiment, it would be possible to determine the entire frequency spectrum of geomagnetic waves.

6.4.1 - Consequence:

existence of resonances ⇔ existence of earth magnetic waves

The physical functionality of the ejPi method is evident by appropriate measurements.
Technical difficulties are here the smallness of the geomagnetic values (< 10 μT) and the technical interference voltages that are larger by a factor of 1000 to 10000.

6.4.2 - Consequence:

Tensions in the earth produce piezoelectric effects. Current or voltage peaks occur during discharge (earthquake) for a short time and are associated with magnetic fields.
Magmatic flows in volcanoes represent also electric charges which produce magnetic fields.
The ejPi-method is passive as well as active for earthquake recognition and measurement, as well as for the detection and measurement of volcanic activities.
Thus, an electromagnetic detection of volcanism and plate tectonics are possible.

6.5 – Synthesis

If earth magnetic waves exist, still following conclusion can be pulled. According to chapter 3.9 applies:

$$H_{k,n} = 255^{k-1} \cdot \frac{4}{5} R_E \cdot e^{(-1)^k \cdot \frac{8n}{15}}$$

with

$$A_k = \frac{4}{5} \cdot 255^{k+1}$$

and

$$q_k = \frac{5}{18} \cdot (-1)^k$$

you can generally write for the radial structure:

$$H_n = A_k \cdot R_E \cdot e^{q_k \cdot n}$$

According to chapter 4.5 applies to the angle part of the field:

$$B = \sum_{i=0}^{n} \sum_{j=0}^{m} \left(a_{ji} \cdot \cos j\varphi + b_{ji} \cdot \sin j\varphi\right) \cdot (\cos i\lambda + \sin i\lambda)$$

Then the following equation can be generally for an earth oscillation structure formulated, that also represents a solution of Laplace's equation:

6.5.1 - Equation:

$$\boxed{S(r,\lambda,\varphi)_{Norm} = A \cdot \frac{r}{R_E} \cdot e^{qn} \cdot \sum_{i=0}^{n} \sum_{j=0}^{m} \left(a_{ji} \cdot \cos j\varphi + b_{ji} \cdot \sin j\varphi\right) \cdot (\cos i\lambda + \sin i\lambda)}$$

This can be understood as an alternative formulation for the equation by Gauß and Weber, because by the equation more the layer and oscillation nature of the field is expressed stronger.

Where against the equations by Gauß and Weber more describe the vector nature of the field.

The earth magnetic field plays a central role in the earth oscillation structure. For one, it has resulted in the crystallization of the geological layers to polyhedron structures in the earth.
On the other hand, it is still involved in maintaining the electrical conductive layers of the atmosphere and thus the electric field of the earth.
This context requires that the magnetic field is a formative factor with in weather and climate formation.

7 – Converting a numeric sequence into an e-function

Given are concentric arrangements such as the layers of the sun, the orbit of the planets, the moons and the rings of planets or like an orange, coconut, dahlia or Narcissus, etc. This chapter shows that concentric arrangements can be presented as exponential or logarithmic functions and are also solution functions of the radial part of Laplace's equation.

Given: a finite ascending or descending sequence of numbers
            (which represent a concentric arrangement)

The evaluation of an e-function from a series of numbers runs in four steps - numbering, logarithmic, linearization and function forming.

7.1 – Numbering

The simplest approach of scale formation consists of numbering the values. On the basis of the following treatments is to keep in mind:

a) same (duplicate) values receive the same number

b) values that differ only slightly from each other justify not a whole counting step in numbering and must be adapted to the count, by shrinking the counting step.
This plays a role in the later linearization. (See example Sun page 140, 141)
Two almost identical values create a staircase in the course of the function, in the later formation of the linear function.
For the smoothing of the function it is necessary that the values in the numbering have to be close to each other.

c) to the later determination of the e-function, it is better than to start counting with zero.

So are **n+1** values, namely: $w_0, w_1, w_2, \ldots w_k, \ldots w_n$
and there is a minimum value w_{min} and a maximum value w_{max}

At least 2 values are needed for the later formation of the minimum-maximum. But at least 3 values are required for the following logarithmic and linearization. Otherwise not a unique approximation line can be made.

Therefore at least three values are required: **n+1 ≥ 3**

7.2 – Logarithmic

The given values w_k are logarithmised (in the next step). The logarithm values are represented as a **function** of the numbering. See the pictures from illustration 7.1.

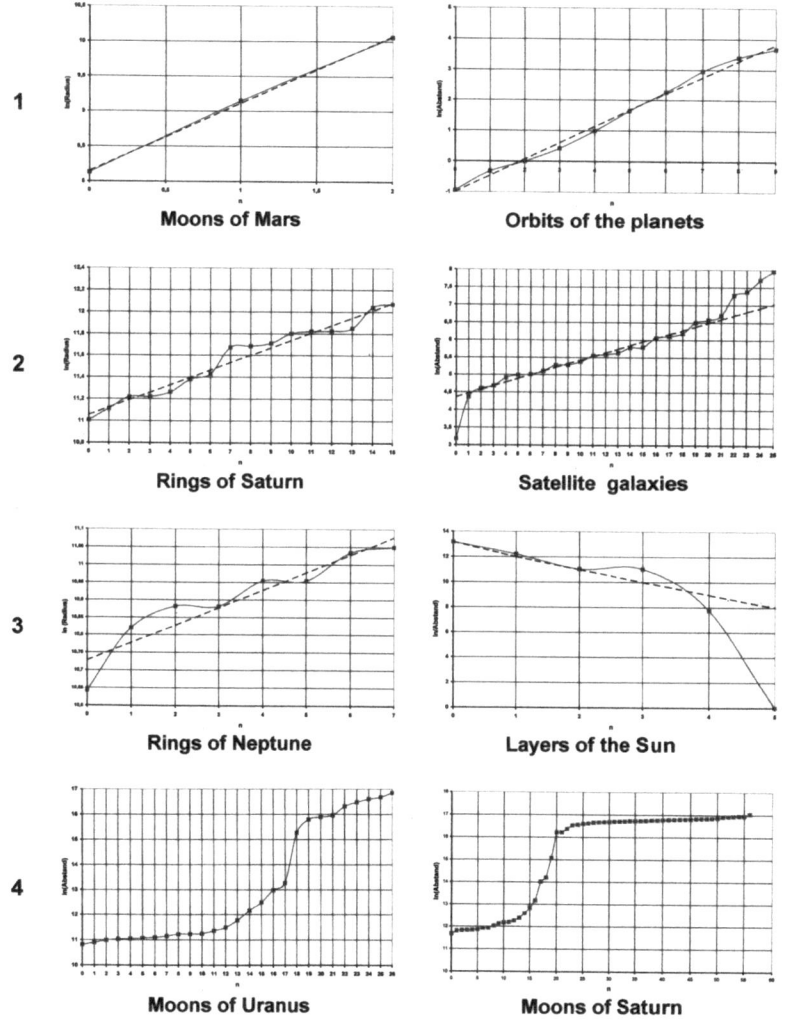

Illustration 7.1 – logarithmic values

There must be at least a nearly linear response function for the logarithmic.
This linearity is a necessary precondition that the sequence n values can be converted in an e-function.
If only two values are present the second value on numbering anywhere can be pushed. Thus, no clear line is definable. Only if there are at least **three** values, a unique function line can be extracted from this.
The illustration 7.1 show the various types of functions, how they can become real at the logarithmic.

Line 1
For the pictures of the moons of Mars and the orbits of the planets is a nearly perfect linearity of the logarithmic values to recognize.

Line 2
For the image of the Saturn rings is a good linearity of the logarithmic values to detect.
For the image of the satellite galaxies is to recognize that most of the data have already a good linearity. Only at close range and in the far range values exist by way of derogation, which can then be smoothed by the following linearization.

Line 3
For the image of the Neptune rings is a good linearity of the logarithmic values to detect.
The layers of the sun look non linear at first glance. You should note that point 3 here roughly at the same height as point 2 is, what does not justify a whole increment. If point 4 is set to point 5, arises a remarkably good linearity. This shows that a greater linearity can be reached via an approximate numbering.

Interim result
The linearity of the logarithmic values in the examples of lines 1, 2 and 3 is already so good that a strong relation to the e-functionc can be assumed here.

Line 4
At the moons of Uranus, the staircase shaped structure is clear to see. The step rise between the points 17 and 18 means nothing more than the existence of a large new space created between the orbits.
The orbits of the moons of all planets can be splitted into two sections. In a **close range** and in a **far range**. It separates the two areas and treated separately, so a good linearization achieved here too.
At the moons of Uranus, the stairway shaped structure is also clear to see. Here, too, a close and a far range arise as with Uranus. In contrast to Uranus, but also moons in the area of the gap exist, so that you can speak here of a **middle range**.

When charts such as the moons of Saturn and Uranus appear, only the data from the close range are used. The middle and far range is then obtained by extrapolation of the found function.

7.3 – Linearization

The values are now linearized to get a possible perfect straight line. The e-function can then directly be determined from this line.
If there is sufficient linear behavior of the function after the logarithmic, (to see later in the Mars moons and peach) a following linearization can be omitted.

If there is still no sufficient linearity, so you can still handle inappropriate values. The values that do not fit, be moved parallel to the x-axis, until they rest on the approximation line. Then it can be read the new numbering right on the x-axis.

Only minimal changes are necessary to obtain full linearity with the samples of Mars moons, orbits of the planets and the rings of Saturn.
Withall other samples the values must make significant jumps in the numbering part to reach linearity.
In this procedure, a **new** (approximate) numbering of the values can arise.
These **new approximate numbering** is then used for further evaluation. This is important for further treatment to achieve largest equivalence for the e-function.

7.4 – Determination of the approximation line

There are two ways to determine a proximity line from the logarithm (and linearized) data.

    a)   by linear regression (without previous linearization)

    b)   by the existing minimum-maximum values

In the following case b) is here used, since the case a) can be treated through commercially available calculation programs.

Wanted is now the approximation line **y = ax + b** for the logarithmic values. In the following illustration, the approximation line is drawn as a dashed line.

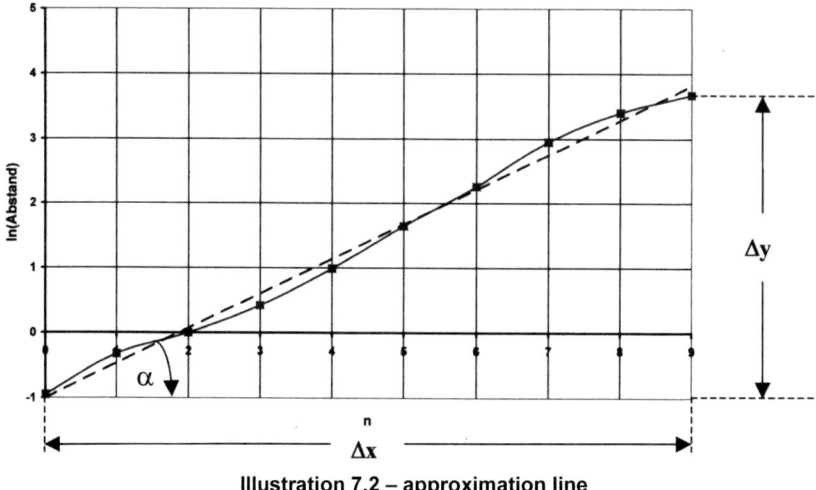

Illustration 7.2 – approximation line

There are **n** values given, namely: $y_0, y_1, y_2, \ldots y_k, \ldots y_n$

with $\qquad y_k = \ln w_k$

There is a minimum y_{min} and a maximum y_{max}

The slope of **a** the approximation linet can be determined from the min-max values and the new approximate numbers. It applies:

7.4.1 - Equation: $\qquad a = \tan \alpha = \dfrac{\Delta y}{\Delta x}$

Δy is the difference between minimum and maximum value:

7.4.2 - Equation: $\quad \Delta y = \ln w_{max} - \ln w_{min} = \ln w_n - \ln w_0$

Δx is the maximum value of the new numbering:

7.4.3 - Equation: $\quad \Delta x = n_{neu}$

The additative constant of this function arises from the smallest value:

7.4.4 - Equation: $\quad b = \ln w_0$

For the approximation line:

7.4.5 - Equation: $\quad \ln w_k = y_k = a \cdot x + b$

Inserting all terms yields:

7.4.6 - Equation: $\quad \boxed{y = \frac{1}{n} \cdot \ln \frac{w_n}{w_0} \cdot x + \ln w_0}$

7.5 – Determination of the e-function

Basis for the formation of the e-function is the linearized function:

$$\ln w_k = y_k = a \cdot x + b$$

The e-function can directly be derived from this linearization:

7.5.1 - Equation: $\quad w_k = w_0 \cdot e^{ax}$

7.6 – Determination of a new numbering

There is the possibility for the numbering values to get an even better fit. Starting from the linearization exact number can be calculated. It applies:

$$\ln w_k = y_k = a \cdot x + b$$

Rearranging the equation for **x** yields:

7.6.1 - Equation:
$$x_k = \frac{\ln w_k - b}{a}$$

Inserting all terms yields:

7.6.2 - Equation:
$$x_k = n_{neu} \cdot \frac{\ln w_k - \ln w_0}{\ln w_n - \ln w_0}$$

With the new numbering, you get a series of numbers on the **basis of logarithms corresponding to the output values**. This new numbering then reflects the harmonical structure of the studied arrangement and can be considered as new scale.

Replaced it in all equations **ln** through **log** and **e** by **10**, so the entire viewing and function discovery applies also to base 10.

7.7 – Global Scaling

There are **n** values given, namely: $w_0, w_1, w_2, \ldots w_k, \ldots w_n$

There are calibration values: M_{eich} (which are based on the physical size of the Protone)

The scaling in the Global Scaling is defined by the following equation:

7.7.1 - Equation: $$S_k = \ln \frac{measuringValue}{scalingValue} = \ln \frac{w_k}{M_{Eich}}$$

Owing to the logarithms rules: $\quad S_k = \ln w_k - \ln M_{eich}$

Applies according to the equation 7.4.5: $\quad \ln w_k = y_k = a \cdot x + b$

So, it can be written: $\quad S_k = y_k - \ln M_{eich}$

Use of the term for y: $\quad S_k = ax_k + b - \ln M_{eich}$

Applies according to the equation 7.4.4: $\quad b = \ln w_0$

In total:

7.7.2 - Equation: $$S_k = a \cdot x_k + \ln \frac{w_0}{M_{eich}}$$

Inserting all terms yields:

7.7.3 - Equation: $$S_k = \frac{\ln w_n - \ln w_0}{n_{neu}} x_k + \ln \frac{w_0}{M_{eich}}$$

If the standard gauge ($M_{eich} = 1$) is set to 1 you simply get the equation 7.4.5:

$$S_k = y_k = ax_k+b$$

This means:

1) It exists a functional relation between the here described e-function determination and the global scaling.

2) The procedure described here provides a form of scaling which is of fundamental nature, while global scaling represents a rather derived size, because a specific standrd gauge is used there.
This is apparent because in the equation 7.7.2 the gauge occurs only in the additative component of the linear equation.
When all values are given, this term represents only an additative constant. But, the actual scaling process takes place in the first term.

You can even represent that:

The additative constant is given a new name:

7.7.4 - Definition:

$$b_{eich} = \ln \frac{w_0}{M_{eich}}$$

Then you can also represent the scaling function:

7.7.5 - Equation:

$$S_k = a \cdot x_k + b_{eich}$$

So here we have a linear equation in front of us. The slope of the line is represented by the first term **ax**. The additative constant **b** moves this line only in vertical position, so along the y-axis.

3) The here described procedure is equivalent to the global scaling. This confirms global scaling again in terms of the logarithmic structure in the universe (see also chapter 8).
However, the equations 7.7.2 and 7.7.4 show that the choice of gauge is **arbitrary**. Instead of the proton variables in the global scaling can be used **any other** size such the electron.

The gauge setting to 1 ($M_{eich} = 1$) is mathematically much more natural to use. Then the harmonical sizes can be determined according to equation 7.6.1 and 7.6.2.
The orbit of the planets are used as an example here. The first numbering (old) is included in the table as well as the approximate numbers of the linearization and the calculated numbers. You can transform the calculated values with equation 7.7.2 and 7.7.4 into the global scaling values.

Planet	Nr alt	distance [AE]	Nr neared	x_k calculated	$a \cdot x_k$	Global Scaling $a \cdot x_k + b_{eich}$
Mercury	0	0,3871	0	0	0	60,88009872
Venus	1	0,723	1,3	1,182	0,624726165	61,50482488
Earth	2	1	1,9	1,796	0,949072221	61,82917094
Mars	3	1,524	2,6	2,593	1,370410679	62,25050940
Asteroids	4	2,7	3,75	3,675	1,942323994	62,82242271
Jupiter	5	5,203	5	4,916	2,598307604	63,47840632
Saturn	6	9,582	6,1	6,071	3,208958560	64,08905728
Uranus	7	19,201	7,5	7,386	3,904034582	64,78413330
Neptune	8	30,047	8,25	8,233	4,351835044	65,23193376
Pluto	9	39,482	8,75	8,750	4,624917093	65,50501581

In the global scaling the self wavelength of the proton with a value of $2,103089 \cdot 10^{-16}$ m is described. Thus arises $b_{eich} = 60,88$.
Applies to the orbits of the planets: $a = 0,52856$ (see page 151)
The calculated values of the numbering x_k reflect the absolute harmonic ratios of a system and are completely independent of a gauge. At the global scaling the actual size of the harmonical $a \cdot x_k$ moves through the constant addend b_{eich} only in the scale. The choice of the calibration parameter is therefore arbitrary.
Also in this chapter you can replace **ln** in all equations with **log**. So, the whole look and function determination applies also to the base 10.

8 – Concentric arrangements

Generally describes the equation 6.5.1 oscillation phenomena around or in a central body.
Stratification or grid formation or polyhedra forming can be considered as a direct expression of an oscillation phenomenon.
The radial part of Laplace's equation applied onto the atomic level, leads to the Schrödinger equation or the orbital model for atoms.
Here the question arises how far this oscillation model can be applied on other spherical and concentric phenomena of our world. To do this, you need only to examine that concentric arrangements could be converted in an e-function.

8.1 – The sun

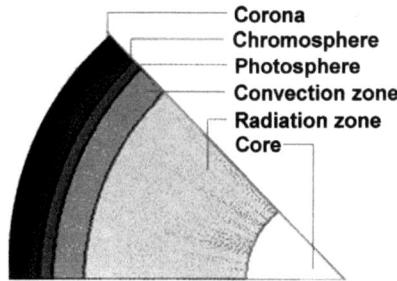

Also, our central star, the sun, features a layer design. Six relevant layers are known in the general literature.

Illustration 8.1.1 – layers of the sun

The individual layers are sorted by depth and then numbered.

Layer	Nr	R (Km)	Depth (Km)	ln(Depth)
core	0	174000	522350	13,16609314
radiation zone	1	494400	201950	12,21577542
convection zone	2	633670	62680	11,0457977
photosphere	3	634100	62250	11,03891381
chromosphere	4	694000	2350	7,762170607
surface	5	696350	0	0

So **n** values are given, namely: $w_0, w_1, w_2, \ldots w_k, \ldots w_n$

The values are sorted by descending size. The logarithm values are represented as a function of the numbering. The following function **y** of the logarithms of the depth is apparent from the table:

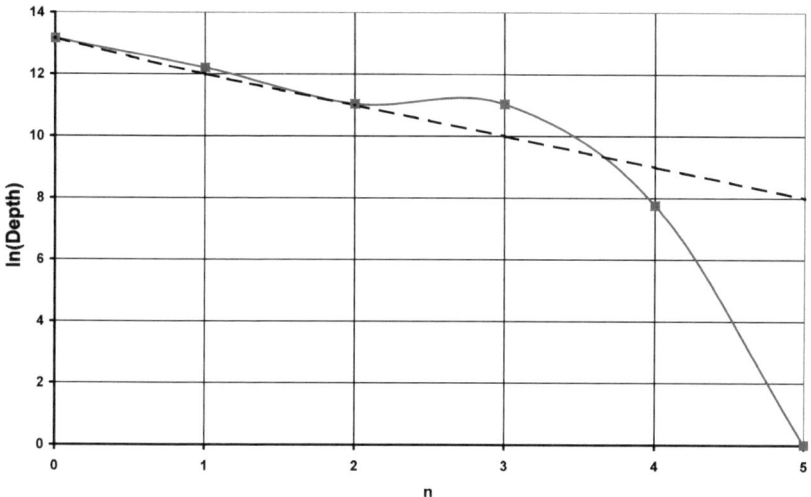

Illustration 8.1.2 – logarithmic layers of the sun

The dashed line in the diagram represents an approximation line. To obtain later an e-function, a straight line as a function must exist at the level of the logarithm. The linearity is a **necessary** precondition that the sequence of **n** values can be converted into an e-function. It is considered that duplicate values get the same number.

The linearity can further be increased if the values (which do not fit yet) are moved parallel to the x-axis, until they rest on the dashed line of proximity. Then you can read the new numbering right on the x-axis.
Is point 3 here is roughly at the same height as point 2 which does not justify a whole increment. Point 3 is moved in the direct vicinity of point 2.
If point 4 is set to point 5, arises a remarkably good linearity. The slope between point 4 and point 5 is so great that the previous point 5 upwards up to 12 can be moved.

The linearized function looks now like this:

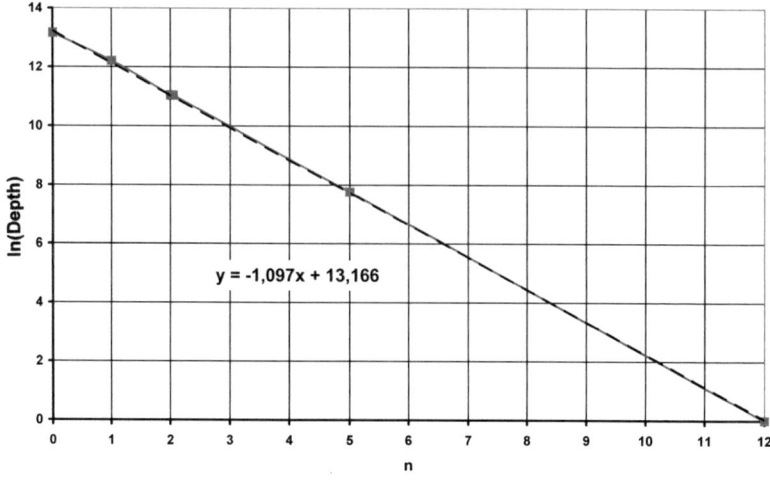

Illustration 8.1.3 – linearization

There should be exist now a **linear** behavior of the generated curve **y**. It is the case as shown in illustration 8.1.3.

The linearity is a **necessary** precondition that the sequence of values can be converted in an e-function. Looking for the approximation line is **y = a·x + b** for the logarithmic values. The line applies the slope **a**:

$$a = \tan \alpha = \frac{\Delta y}{\Delta x}$$

Because a sorted sequence of values exists, there is a minimum value and a maximum value, namely the first and the last value. Δ**y** can be formed by the difference of the two extremes:

It applies $\quad \Delta y = \ln w_{max} - \ln w_{min} = \ln w_n - \ln w_0$

For the sun layers applies: Δy = 0 - 13,166 = -13,166

Because the x-axis is formed by the numbering, Δ**x** is equal to the number of (new) values:

It applies $\quad \Delta x = n_{neu}$

The sun layer is then: n = 12

The slope a of the line is: a = -13,166/12 = -1,097

Applies to the constant factor **b**: $b = \ln w_0$

For the sun layer is then: b = 13,166

The equation for the approximation line is: **y = -1,0972·x + 13,166**

The e-function can directly be derived from the linearization.

As a general rule: $w_k = w_0 \cdot e^{ax}$

The sun layer is then: **Depth = 522350·e$^{-1,0972 \cdot x}$**

The overall situation for the layers of the Sun looks like this:

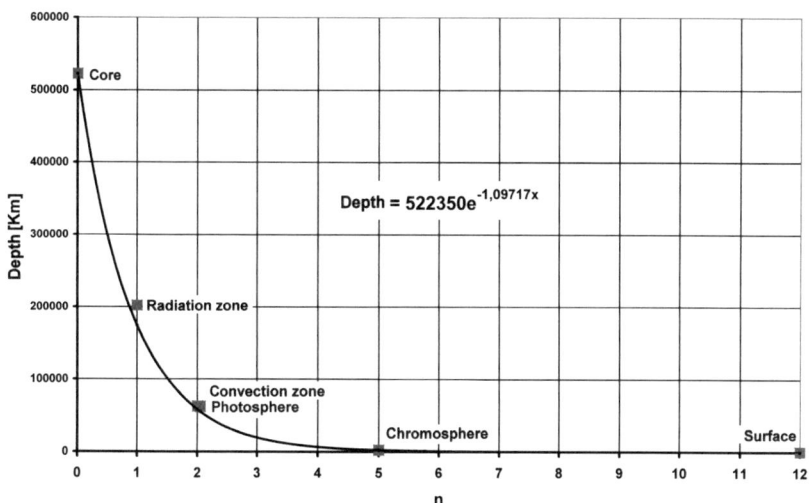

Illustration 8.1.4 – layers of the sun as an e-function

There is still the possibility for the numbering values to get an even better fit. Starting from the existing linearisation accurate numbers can be namely calculated.

It applies:
$$\ln w_k = y_k = a \cdot x + b$$

Rearranging the equation to x yields:
$$x_k = \frac{\ln w_k - b}{a}$$

Inserting all termes yields:
$$x_k = n_{neu} \cdot \frac{\ln w_k - \ln w_0}{\ln w_n - \ln w_0}$$

The following table shows the calculated values of the numbering as well as the approximate numbers.

Layer	Depth [Km]	Nr calculated	Nr neared
core	522350	0	0
radiation zone	201950	0,866	0,9
convection zone	62680	1,932	1,9
photosphere	62250	1,938	2
chromosphere	2350	4,925	5
surface	0	12	12

The more accurate function for the sun layers is in illustration to see 8.1.5. Adding still the center of the sun, the situation is changing only minimally. The next table and illustration 8.1.6 represent the situation.

Layer	Depth (Km)	Nr calculated	Nr neared
sun center	696350	0	0
core	522350	0,2564498	0,25
radiation zone	201950	1,1040895	1,1
convection zone	62680	2,1476559	2,2
photosphere	62250	2,153796	2,25
chromosphere	2350	5,0765004	5,1
surface	0	12	12

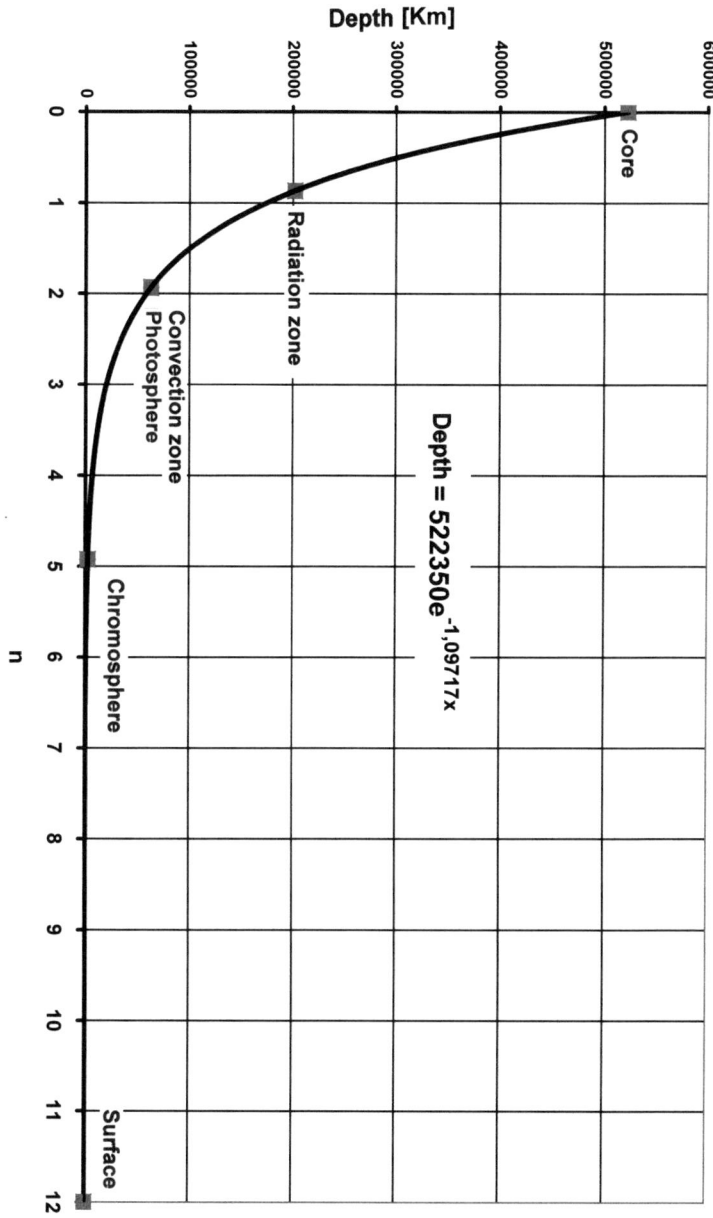

Illustration 8.1.5 – layers of the sun

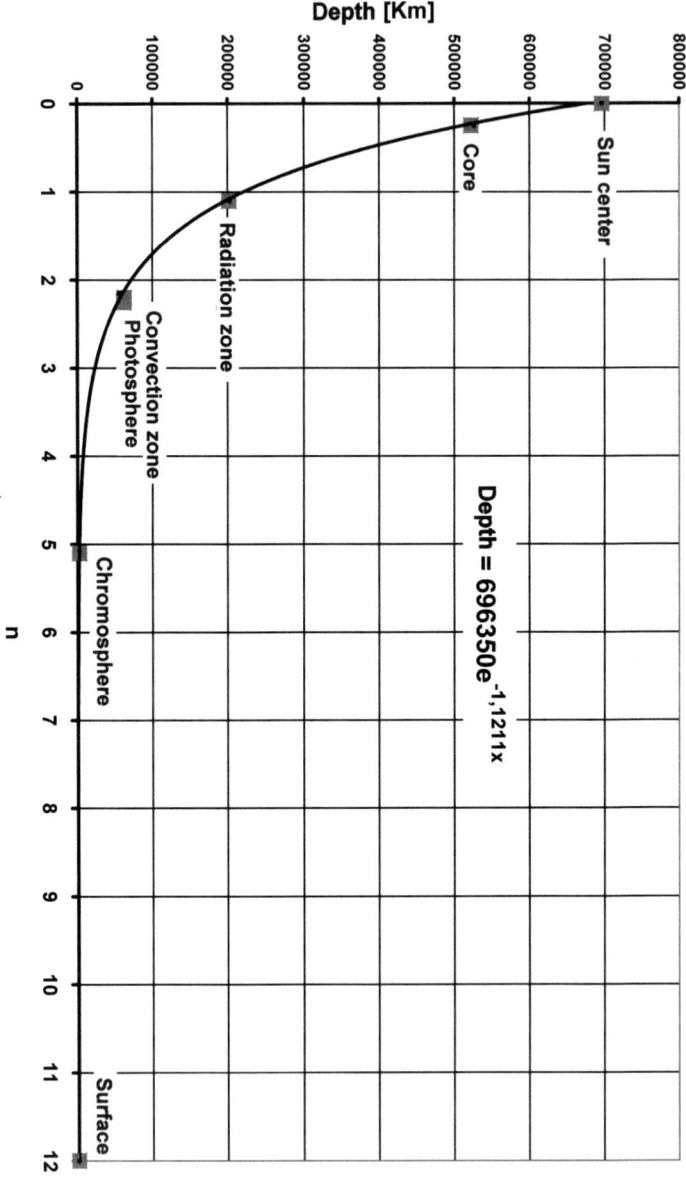

Illustration 8.1.6 – layers of the sun

8.2 – The orbits of the planets

Our solar system with the orbits of the planets also has a concentric structure. It carried the distances in astronomical units for the first 8 planets from the popular literature together. Here, the individual planets are assigned according to distance and then numbered.

Planet	Nr	distance (AU)	ln(distance)
Mercury	0	0,3871	-0,94907222
Venus	1	0,723	-0,32434606
Earth	2	1	0
Mars	3	1,524	0,42133846
Jupiter	4	5,203	1,64923538
Saturn	5	9,582	2,25988634
Uranus	6	19,201	2,95496236
Neptune	7	30,047	3,40276282

AU = astronomical unit = 149.597.870 m
The following function **y** of the logarithms of the distances is apparent from the table:

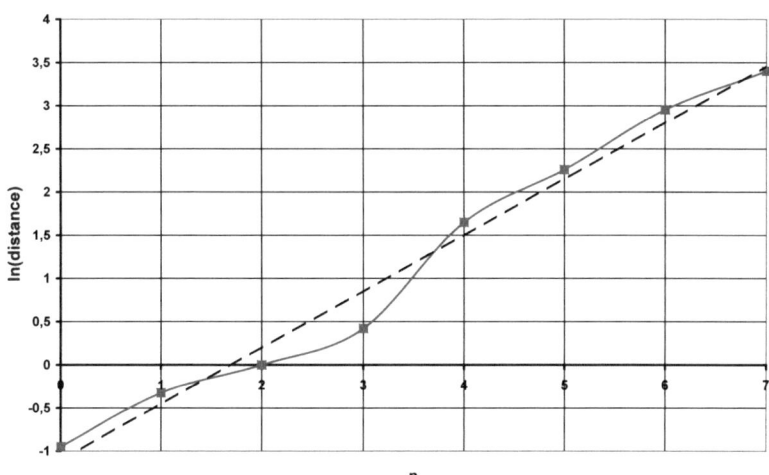

Illustration 8.2.1 – logarithmic orbits of the planets

The dashed line in the diagram represents an approximation line.

How to see a good linearity of the values already exists. Only Earth and Mars vary from the approximation line. Point 3 (Earth) is pushed to point 1.5 and point 3 (Mars) to point 2. The linearized function looks now like this:

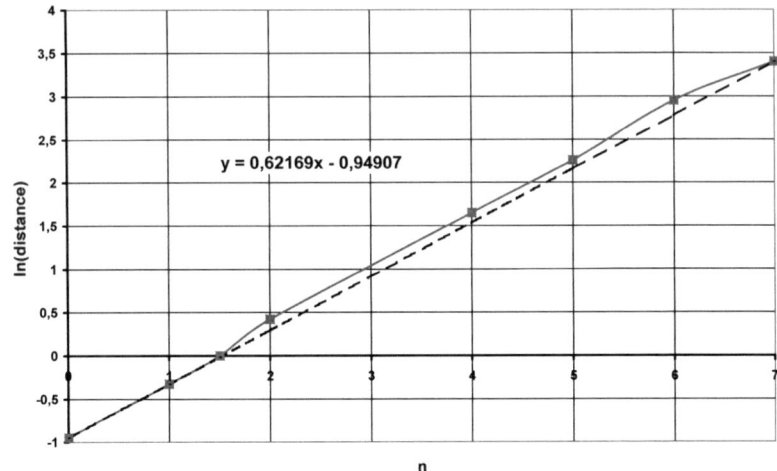

Illustration 8.2.2 – linearization

Described with the same procedure as for the sun layers, the parameters of the approximation straight are determined.

The equation for the approximation line is: **y = 0,62169·x − 0,94907**

Applies to the orbits of the planets: **distance = 0,3871·e$^{0,62169·x}$**

Consequently, the following values appear:

Planet	Nr	distance [AU]	distance [AU]
			calculated
Mercury	0	0,3871	0,3871
Venus	1	0,723	0,7208087
Earth	1,5	1	0,98359983
Mars	2	1,524	1,34219887
Jupiter	4	5,203	4,65383058
Saturn	5	9,582	8,66577519
Uranus	6	19,201	16,1363114
Neptune	7	30,047	30,047

The overall situation for the orbits of the planets looks like this:

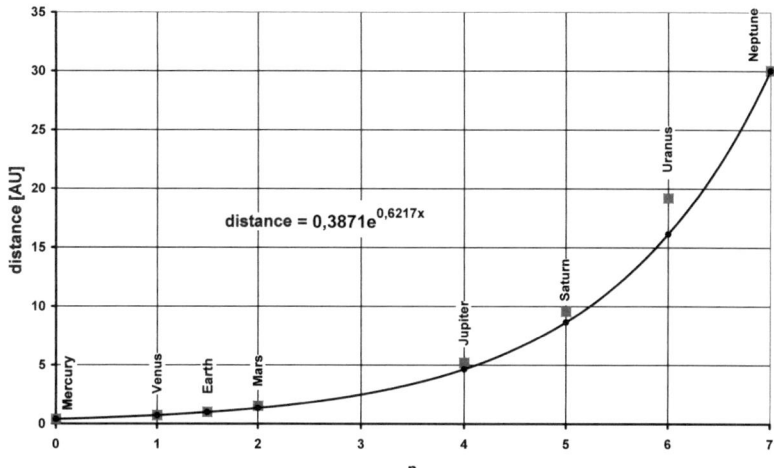

Illustration 8.2.3 – orbits of the planets as an e-function

Remarkable is the gap for n = 3. The distance value is 2.5 AU. The asteroid belt extends from 2.0 to 3.4 AU. The average is 2.7 AE. Thus the model covers very well with the real situation. And it shows that the asteroid belt belongs to the oscillation system of the planets. And it shows also that the procedure of the logarithmic and linearization is not a random act, but revealed the **harmonic** structures of the system.

Therefore a new approach is generated here, this time with the asteroid belt and Pluto.

Planet	Nr	distance (AU)	ln(distance)
Mercury	0	0,3871	-0,94907222
Venus	1	0,723	-0,32434606
Earth	2	1	0
Mars	3	1,524	0,42133846
Asteroid belt	4	2,7	0,99325177
Jupiter	5	5,203	1,64923538
Saturn	6	9,582	2,25988634
Uranus	7	19,201	2,95496236
Neptune	8	30,047	3,40276282
Pluto	9	39,482	3,67584487

There appears the following function for the logarithm of the distance from the table:

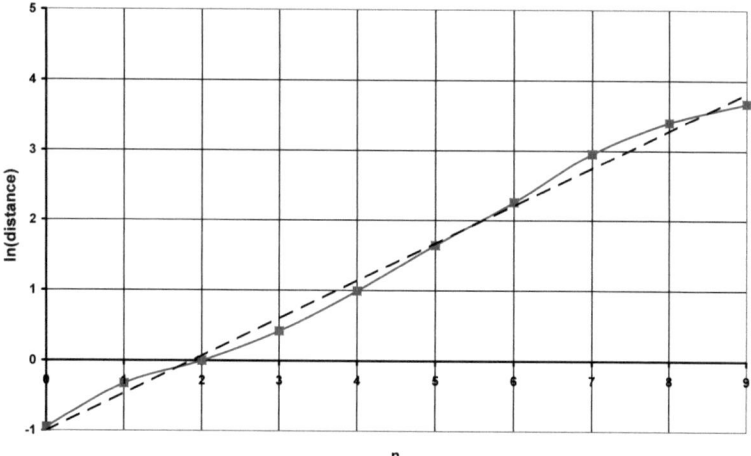

Illustration 8.2.4 – logarithmic orbits of the planets

The dashed line in the diagram represents a proximity line again. As can be seen, there is already a very good linearity response of the values. The values are still linearized and result to the following diagram:

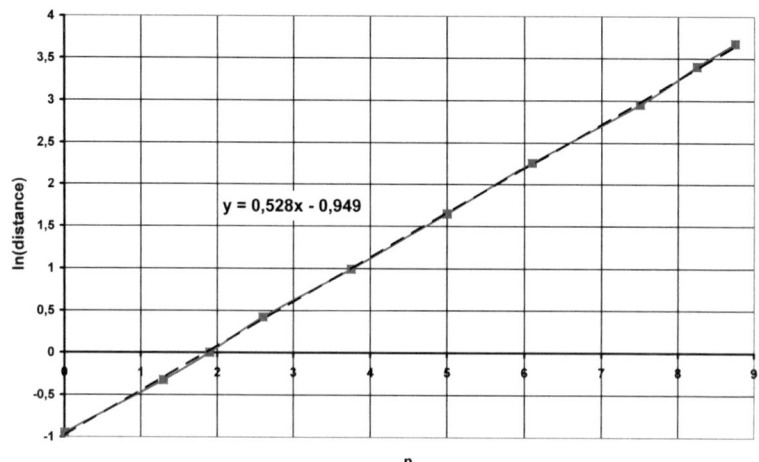

Illustration 8.2.5 – linearization

The equation for the approximation line is: **y = 0,52856·x - 0,94907**

Applies to the orbits of the planets: **distance = $0{,}3871 \cdot e^{0{,}52856 \cdot x}$**

Consequently, the following values will be:

Planet	Nr new	distance [AU]
Mercury	0	0,3871
Venus	1,3	0,723
Earth	1,9	1
Mars	2,6	1,524
Asteroid belt	3,75	2,7
Jupiter	5	5,203
Saturn	6,1	9,582
Uranus	7,5	19,201
Neptune	8,25	30,047
Pluto	8,75	39,482

The new situation for the orbits of the planets looks like this:

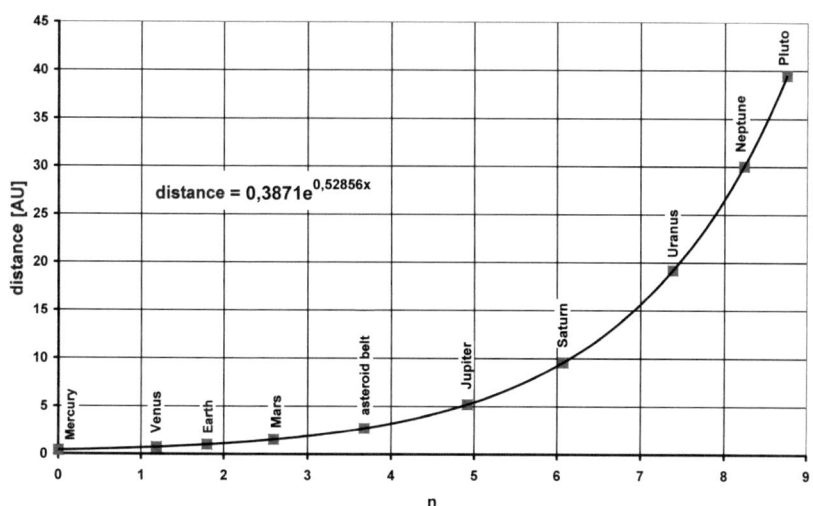

Illustration 8.2.6 – orbits of the planets as an e-function

There is still the possibility to get a better fit for the numbering values. Starting from the existing linearisation accurate numbers can be namely calculated exactly, how it was seen in the sun layers. In the following table are the calculated values of the numbering as well as the approximate numbers and the old numbers listed.

Planet	Nr old	distance [AU]	Nr neared	Nr calculated
Mercury	0	0,3871	0	0
Venus	1	0,723	1,3	1,182
Earth	2	1	1,9	1,796
Mars	3	1,524	2,6	2,593
Asteroid belt	4	2,7	3,75	3,675
Jupiter	5	5,203	5	4,916
Saturn	6	9,582	6,1	6,071
Uranus	7	19,201	7,5	7,386
Neptune	8	30,047	8,25	8,233
Pluto	9	39,482	8,75	8,750

The precise function for the orbits of the planets is shown in illustration 8.2.7. Between orbit radii and orbital periods, a mathematical relationship for the planet exists, that allows to represent also the orbital periods as an e-function.

It is generally:
$$\frac{T^2}{r^3} = C$$

For the mercury is then:
$$\frac{T^2}{r^3} = \frac{T_0^2}{r_0^3} = C$$

By rearranging the equation for the orbital periods of the planets is as follows:
$$T^2 = \left(\frac{r}{r_0}\right)^3 \cdot T_0^2$$

Insertion of all sizes provides:
$$T = T_0 \cdot e^{\frac{3}{2} \cdot a \cdot x} = 0{,}24084 \cdot e^{2{,}02856 x}$$

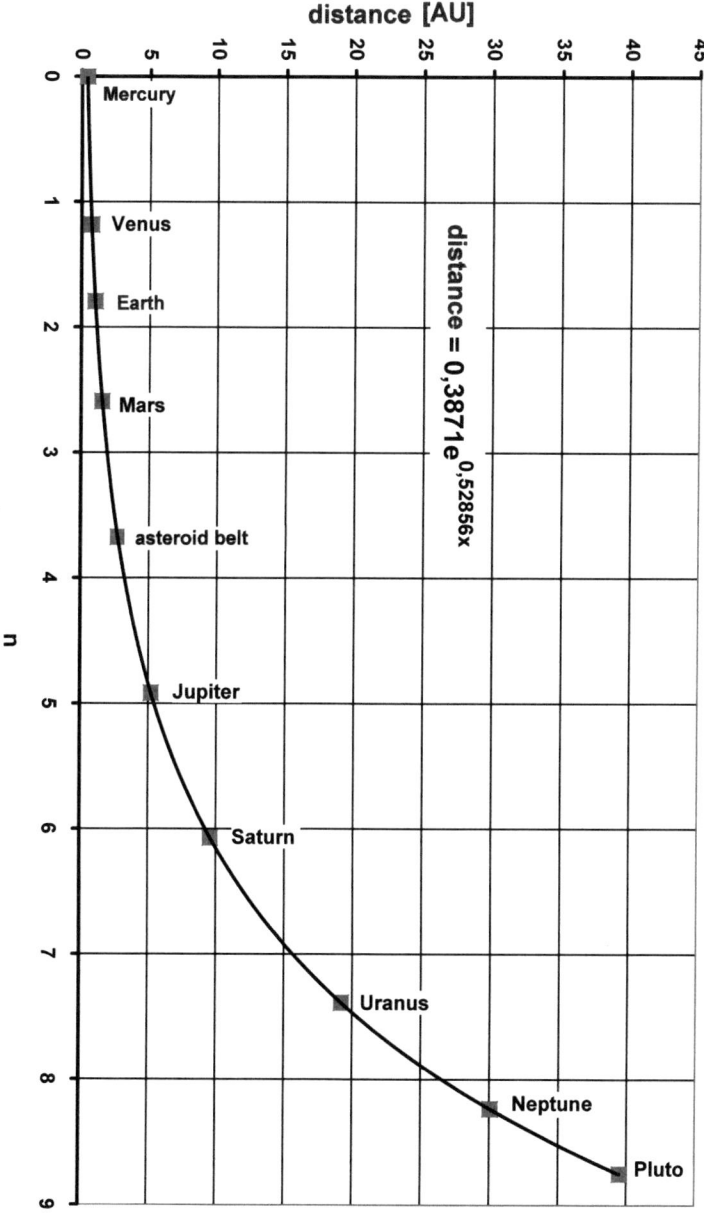

Illustration 8.2.7 – orbits of the planets

8.3 – Moons of the planets

8.3.1 – The moons of Jupiter, Saturn, Uranus and Neptune

In our solar system, the planets have moons, that orbit their planet respectively also on concentric tracks. Considered here are the moons of Jupiter, Saturn, Uranus, Neptune and Mars. Thereby, the logarithmic distances show a slightly different behavior than in the previous examples. The moons of Uranus are shown here for all of these planets.

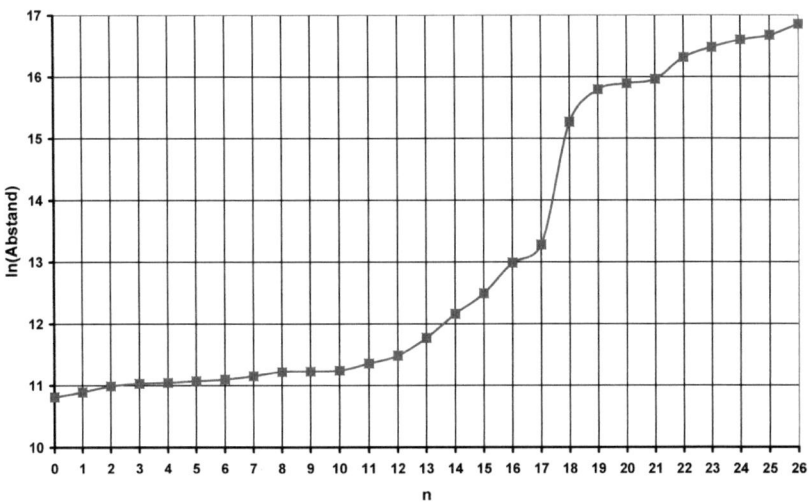

Illustration 8.3.1 – moons of Uranus

The staircase shaped structure is clearly visible. The slope rise between the points 17 and 18 means nothing more than the existence of a large gap between the orbits.
For all planet the orbits of the moons can be split into two sections. In a **close range** and in a **far range**.
To evaluate first of all the values from close range. It can be made then the usual procedure of linearization and functional forming, so as it has been practiced in the previous examples.
Then, if you apply the found function on the far range and so determinating the complete function. On the following pages are the diagrams for the moons of Jupiter, Saturn and Uranus to see.

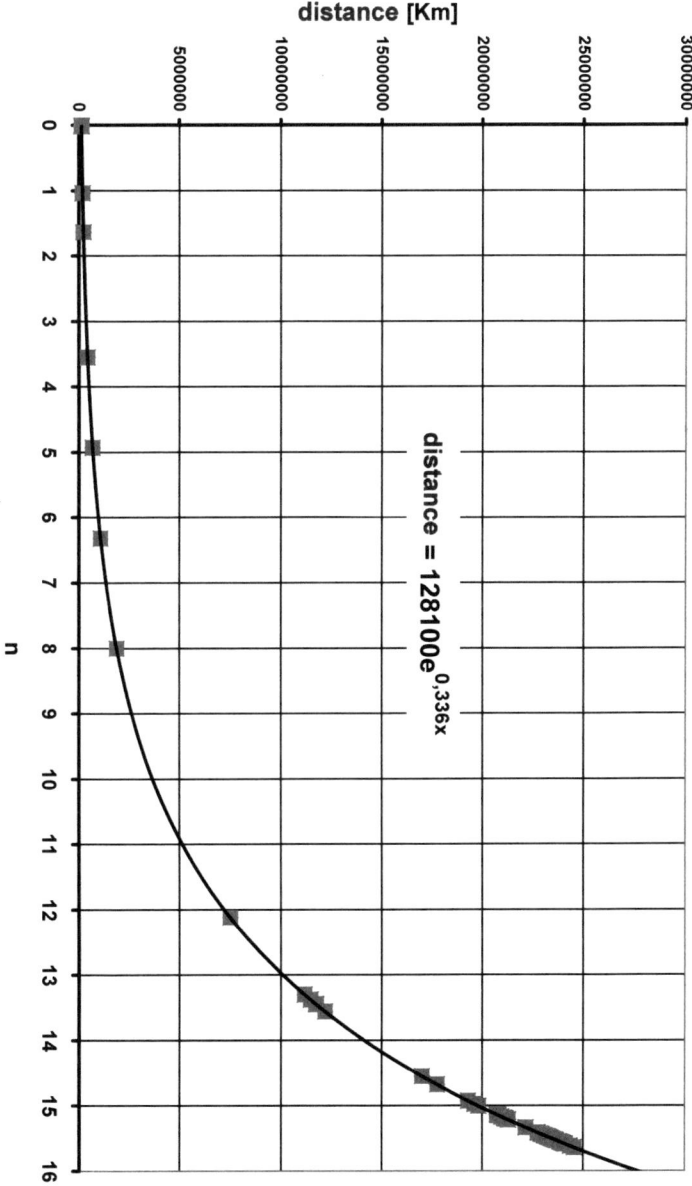

Illustration 8.3.2 – moons of Jupiter

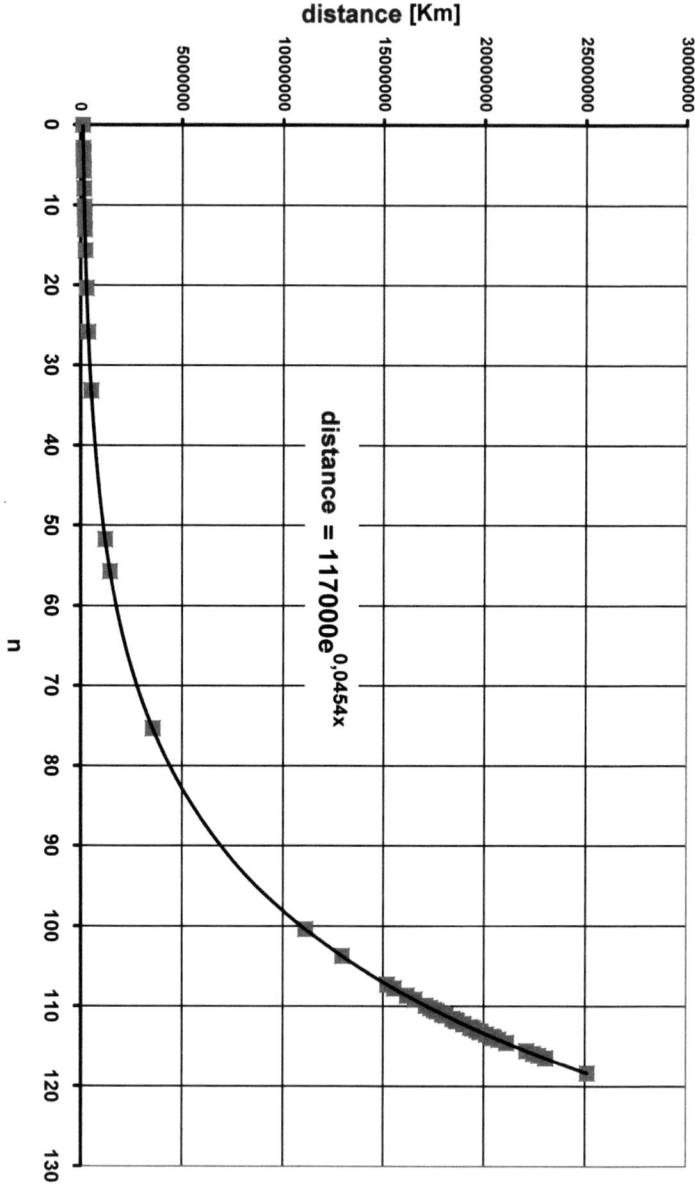

Illustration 8.3.3 – moons of Saturn

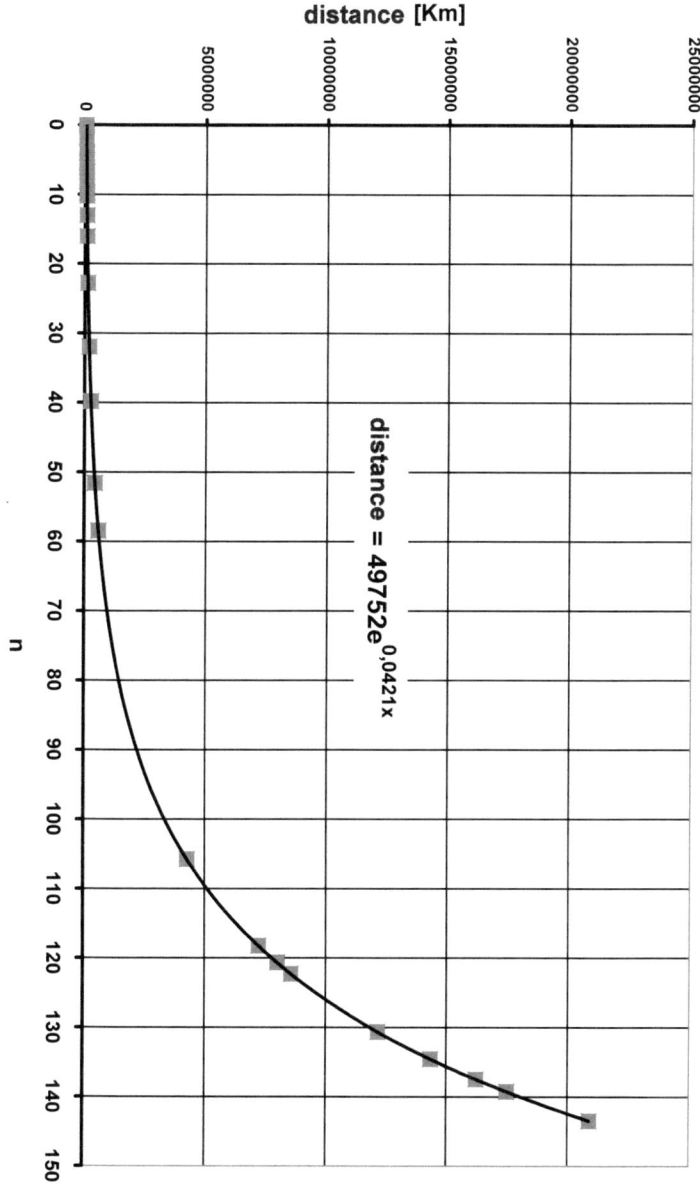

Illustration 8.3.4 – moons of Uranus

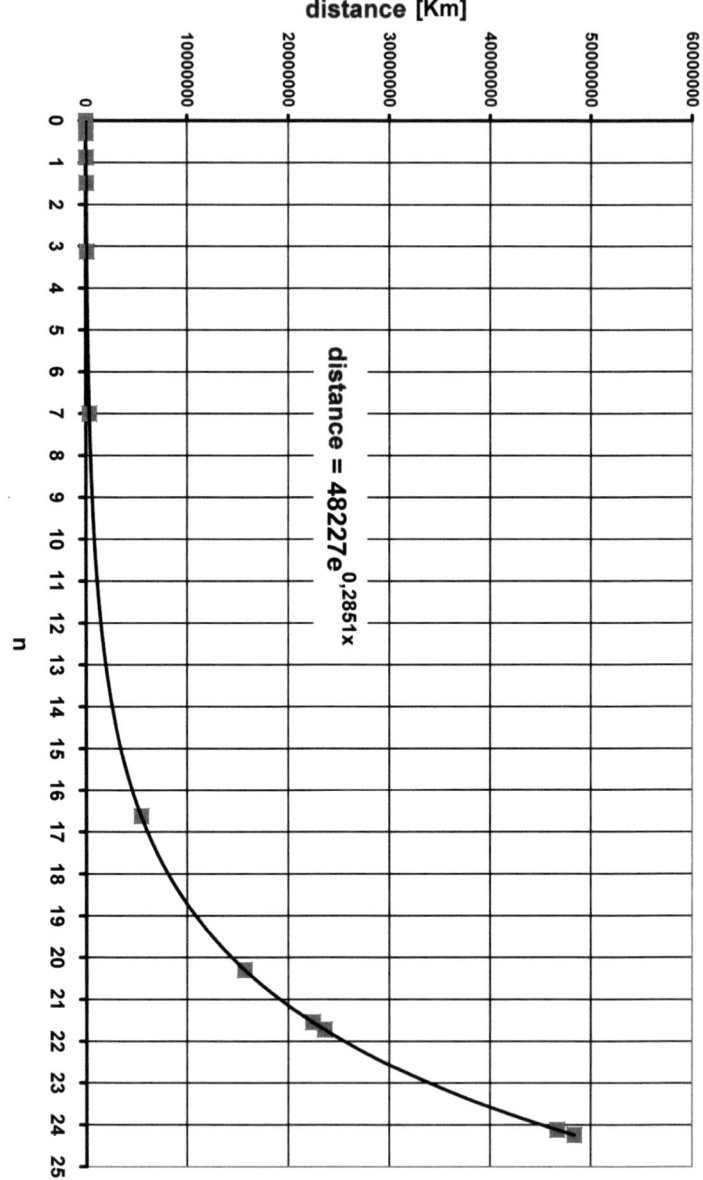

Illustration 8.3.5 – moons of Neptune

8.3.2 – Moons of Mars

3 data points are needed to clearly create an e-function. Mars has only 2 moons. As both moons orbiting Mars quite close, you can take the Mars surface as the third value. The entire issue is again given in the following table.

Name	Radius	Nr	ln(Radius)	Nr calculated
Mars equator	3396	0	8,13035355	0
Phobos	9376	1	9,14590851	1,05
Deimos	23459	2	10,0630095	2

By turning the values into logarithmic numbers and transfer the data into a chart, you get a surprising simplification. The logarithm values are already so linear that a more linearization is no longer necessary.

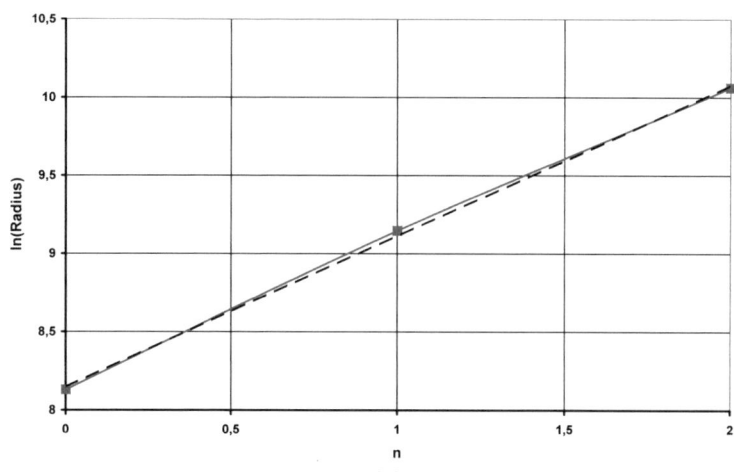

Illustration 8.3.6 – linearization

It can be determined the e-function. The entire situation is shown in the following illustration 8.3.7.

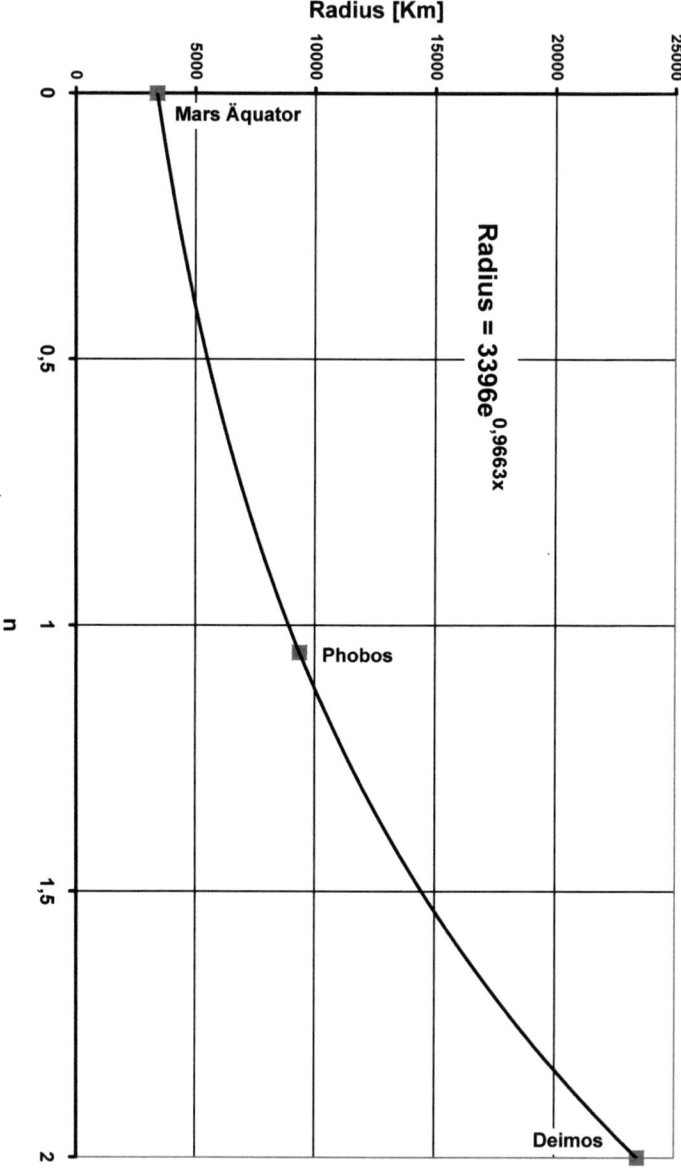

Illustration 8.3.7 – moons of Mars

8.4 – Rings of the planets

More concentric structures in our solar system are the rings of the planets. These include Jupiter, Saturn, Uranus and Neptune.

8.4.1 – Rings of Saturn

The following table shows all the rings of Saturn and the logarithms of the radii.

Name	distance	Nr	ln(distance)
Saturn equator	60.268	0	11,0065566
inside edge ring D	66.900	1	11,1109542
outside edge ring D	74.510	2	11,2186886
inside edge ring C	74.658	3	11,220673
Titan ringlet	77.871	4	11,2628089
Maxwell division/ring	87.491	5	11,3792912
o.e. ring C - i.e. ring B	92.000	6	11,4295439
outside edge ring B	117.580	7	11,6748742
middle Cassini division	119.000	8	11,6868788
inside edge ring A	122.170	9	11,7131688
Encke division	133.589	10	11,8025232
Kepler division	136.530	11	11,8242996
outside edge ring A	136.775	12	11,8260925
middle ring F	140.180	13	11,8506826
inside edge ring G	170.000	14	12,0435537
outside edge ring G	175.000	15	12,0725413
inside edge ring E	181.000	16	12,1062523
outside edge ring E	483.000	17	13,0877719
Phoebe ringlet	12.000.000	18	16,3004172

A close range and a far range like at the moons of the planets also occur whith the Saturn rings.
The evaluation is only taken for the first 16 data points in the close range distance. As usual are the logarithms of the radii in the numbering function applied as in illustration 8.4.1 to see, there is already an approximate linearity of the data. A more linearization of the data is therefore relatively easy to do. The linearized function is depicted in illustration 8.4.2.

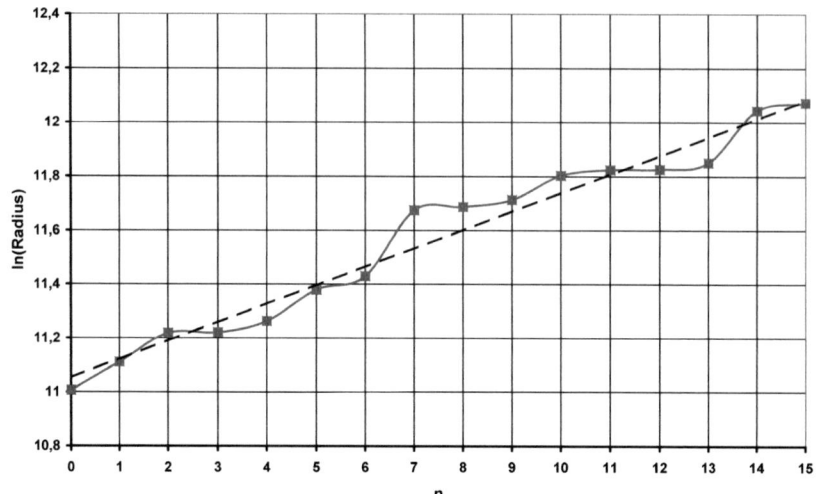

Illustration 8.4.1 – logarithmic radii of the first 16 rings of Saturn

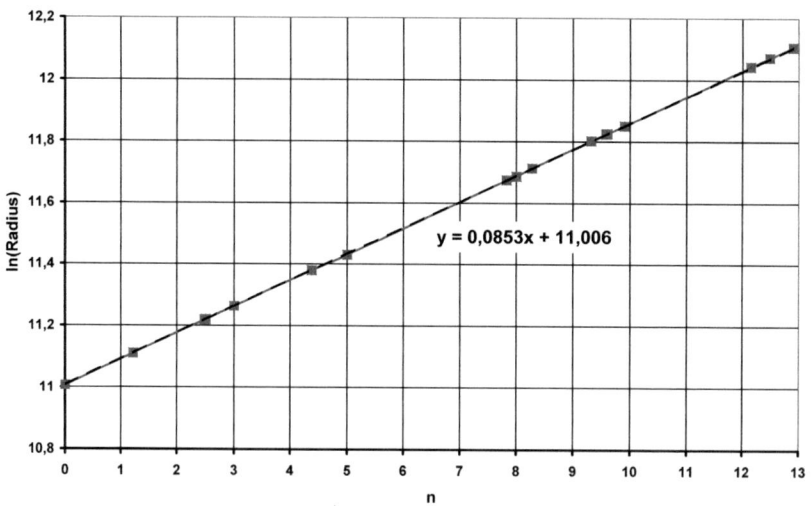

Illustration 8.4.2 – linearization

Also new numbering values can be calculated after the linearization. All numbers are listed in the following table.

Name	distance	Nr old	Nr neared	Nr calculated
Saturn equator	60.268	0	0	0
inside edge ring D	66.900	1	1	1,22
outside edge ring D	74.510	2	2	2,48
inside edge ring C	74.658	3	2,1	2,5
Titan Ringlet	77.871	4	2,5	3
Maxwell division /ring	87.491	5	4	4,37
i.e ring C - o.e. ring B	92.000	6	4,5	5
outside edge ring B	117.580	7	7,5	7,83
middle Cassini division	119.000	8	7,75	8
inside edge ring A	122.170	9	8	8,28
Encke division	133.589	10	9	9,32
Kepler division	136.530	11	9,2	9,58
outside edge ring A	136.775	12	9,3	9,6
middle ring F	140.180	13	9,5	9,9
inside edge ring G	170.000	14	11,75	12,14
outside edge ring G	175.000	15	12	12,48
inside edge ring E	181.000	16	12,5	12,9

The e-function can again be determined from the linearized values. The rings of Saturn in the close range are shown in illustration 8.4.3. The illustration 8.4.4 shows the entire facts for Saturn.

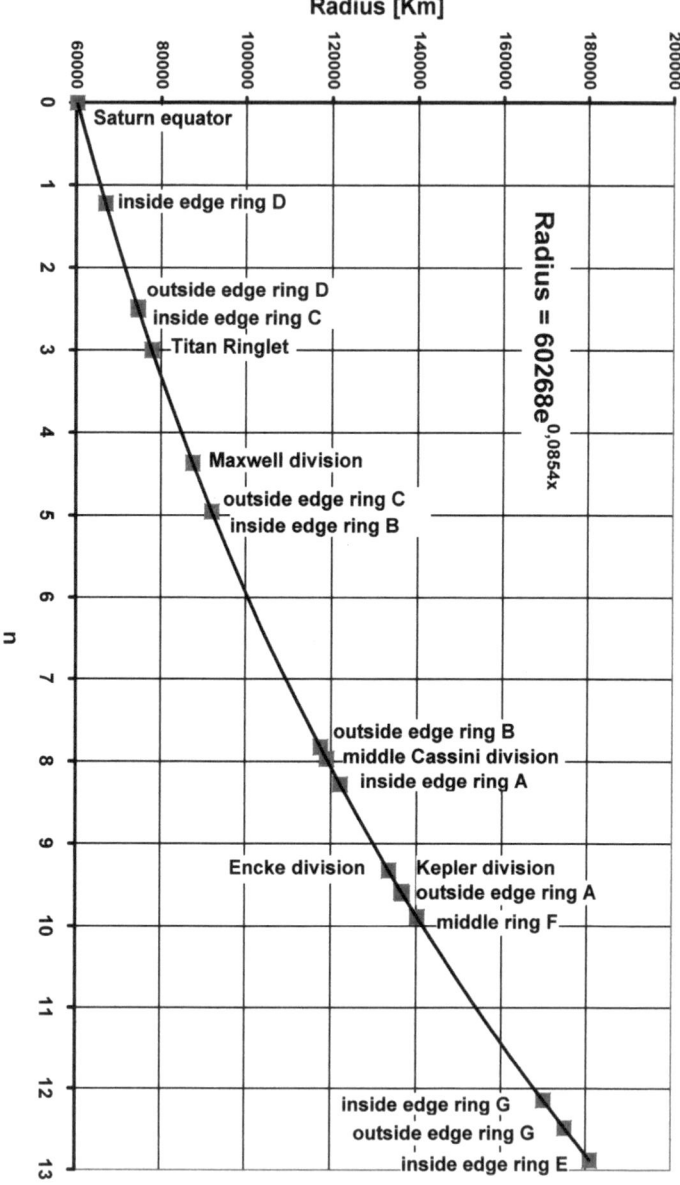

Illustration 8.4.3 – the first 16 rings of Saturn

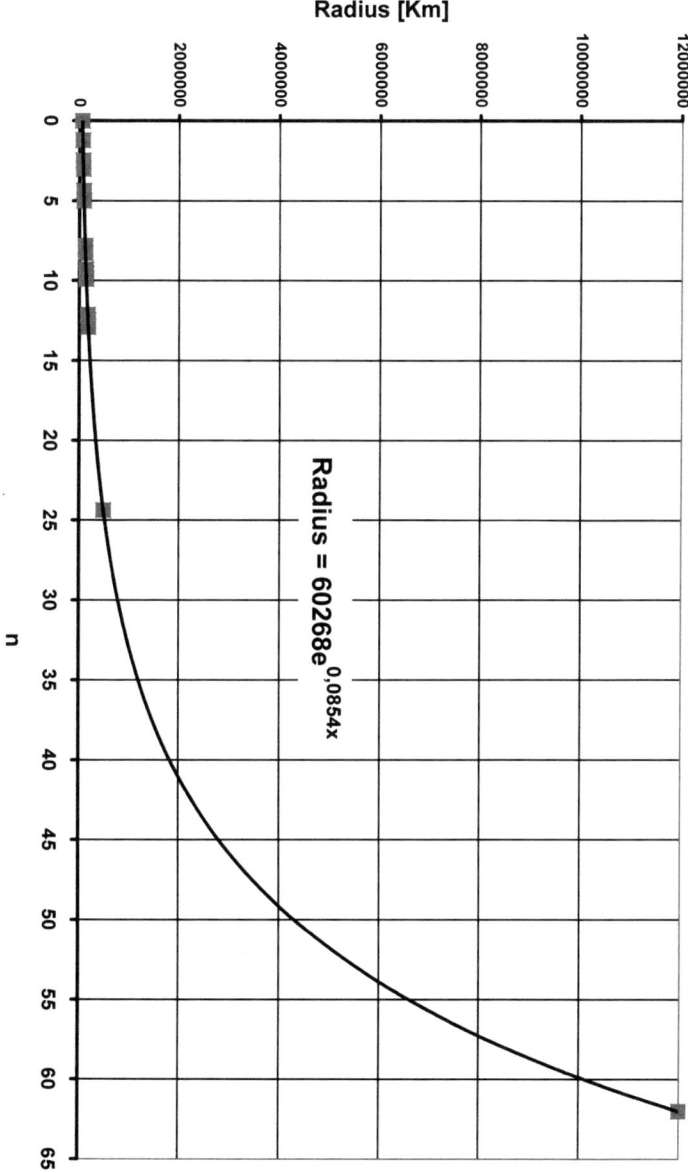

Illustration 8.4.4 – rings of Saturn

8.4.2 – Rings of Jupiter

The following table shows all the rings of Jupiter and the logarithms of the radius, as well as the calculated numbers.

Name	Radius	Nr	ln(Radius)	Nr
	(km)	old		calculated
Halo ring	92000	0	11,429544	0
Halo ring	122500	1	11,715866	0,96
Main ring	122500	2	11,715866	0,96
Main ring	129000	3	11,767568	1,13
Amalthea Gossamer ring	129000	4	11,767568	1,13
Thebe Gossamer ring	129000	5	11,767568	1,13
Amalthea Gossamer ring	182000	6	12,111762	2,28
Thebe Gossamer ring	226000	7	12,32829	3

The e-function can again be determined from the linearized values. The rings of Jupiter are shown in illustration 8.4.5.

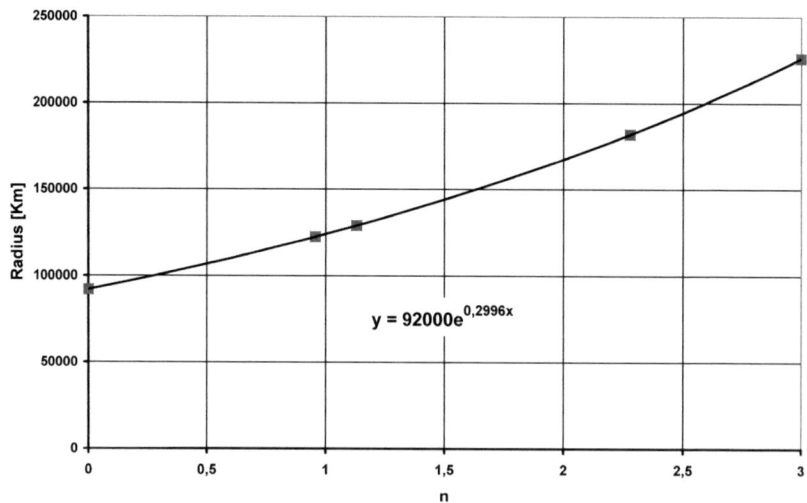

$y = 92000e^{0,2996x}$

Illustration 8.4.5 – rings of Jupiter

8.4.3 – Rings of Neptune

The following table shows all the rings of Neptune and the logarithms of the radius, as well as the calculated numbering.

Name	Radius (km)	Nr	ln(Abstand)	Nr calculated
Galle	41900	0	10,6430411	0
unknown	50000	1	10,8197783	2,28
LeVerrier	53200	2	10,8818137	3,08
Lassell	53200	3	10,8818137	3,08
Lassell	57200	4	10,9543092	4,02
Arago	57200	5	10,9543092	4,02
Not named	61950	6	11,0340829	5,05
Adams	62933	7	11,0498259	5,25

The e-function can again be determined from the linearized values. The rings of Neptune are presented in illustration 8.4.6.

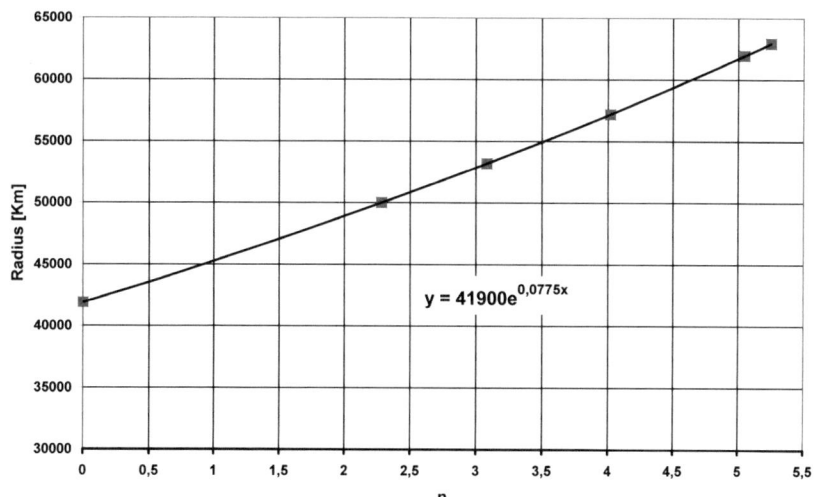

Illustration 8.4.6 – rings of Neptune

8.4.4 – Rings of Uranus

The following table contains all rings of Uranus and the logarithms of the radius, as well as the calculated numbers.

Name	Radius	Nr	ln(Abstand)	Nr
	(km)	old		calculated
ζcc	32000	0	10,3734912	0
1986U2R	37000	1	10,5186732	2,60
ζc	37850	2	10,5413863	3,01
1986U2R	39500	3	10,584056	3,77
ζ	41350	4	10,6298277	4,59
6	41837	5	10,6415364	4,80
5	42234	6	10,6509809	4,97
4	42570	7	10,6589051	5,11
α	44718	8	10,7081314	5,99
β	45661	9	10,7289998	6,37
η	47175	10	10,7616194	6,95
ηc	47176	11	10,7616406	6,95
γ	47627	12	10,7711551	7,12
δc	48300	13	10,7851868	7,37
δ	48300	14	10,7851868	7,37
λ	50023	15	10,8202382	8
ε	51149	16	10,8424982	8,40
ν	66100	17	11,098924	12,99
ν	69900	18	11,1548209	13,99
μ	86000	19	11,3621026	17,70
μ	103000	20	11,5424843	20,93

The e-function can again be determined from the linearized values. The rings of Uranus are shown in illustration 8.4.7.

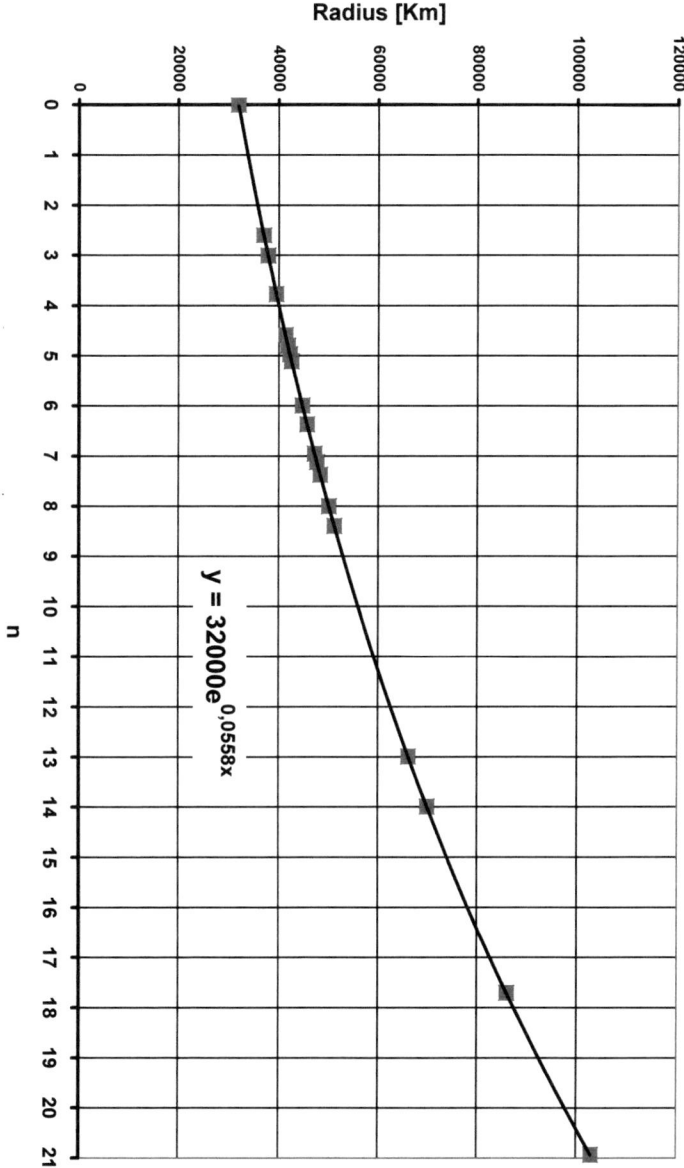

Illustration 8.4.7 – rings of Uranus

8.4.5 – Rings of Rhea

The Saturn moon Rhea also has rings. The following table contains all rings of Rhea and the logarithms of the radius, as well as the calculated numbers.

Name	Radius	Nr	ln(Radius)	Nr
	(km)	old		calculated
1	1615	0	7,38709024	0
2	1800	1	7,49554194	1,17
3	2020	2	7,61085279	2,42
disc	5900	3	8,68270763	14

The e-function can again be determined from the linearized values. The rings of Rhea are shown in illustration 8.4.8.

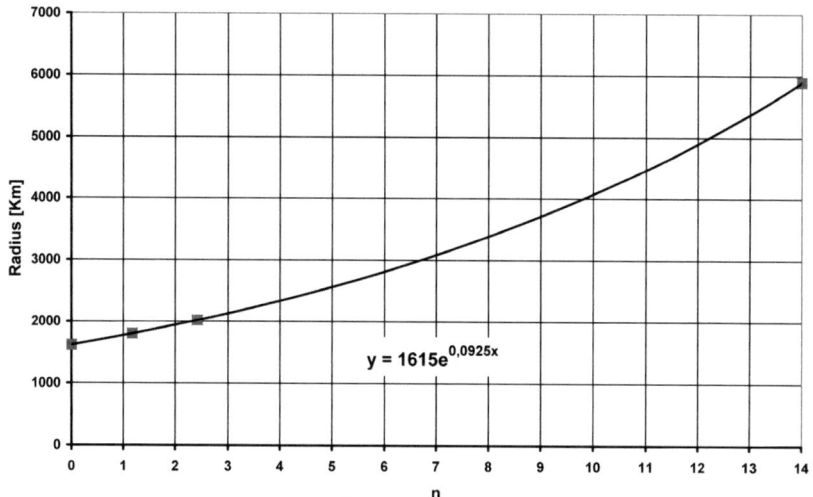

Illustration 8.4.8 – rings of Rhea

8.5 – Satellite galaxies of the milky way

The galaxies that accompanied our Milky Way Galaxy have also a concentric order around our Galaxy.

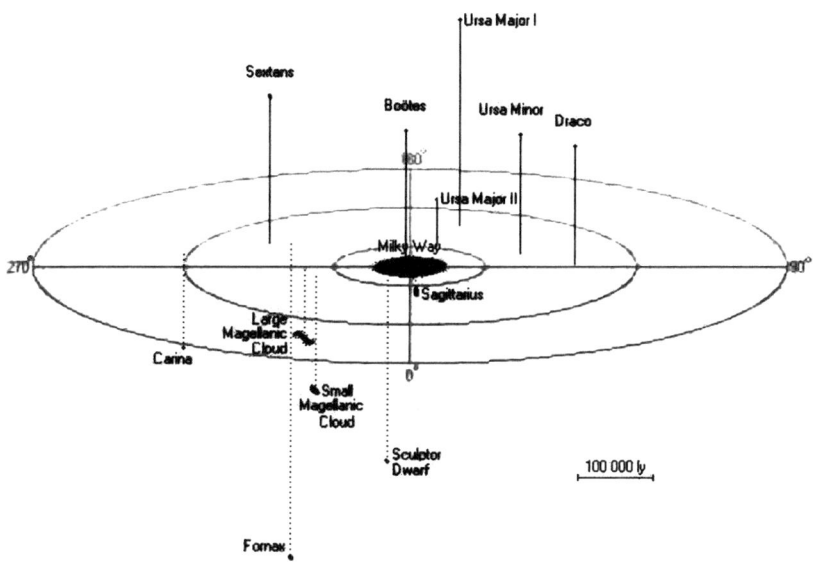

Illustration 8.5.1 – satellite galaxies of the milky way

The following table contains all objects which are generally considered as satellite galaxies. If you look on the tolerances, a part of the distances are provided with major inaccuracies.
The logarithm values of the distances are shown again as a function of the numbers, what is shown in illustration 8.5.2. How to see already a good linearity of the most values exists.

Galaxy	Nr	distance	tolerance	ln(distance)
		light years	light years	
Canis-Major-Dwarf	0	24		3,17805383
Elliptische Sagittarius-Dwarf Galaxy	1	78	±7	4,35670883
Ursa-Major-II	2	100	±15	4,60517019
Complex H	3	108		4,68213123
Bootes-II-Dwarf	4	136	±26	4,91265489
Willman 1	5	147		4,99043259
Bootes-III-Dwarf	6	150		5,01063529
Large Magellanic Cloud	7	165	±5	5,10594547
Small Magellanic Cloud	8	195	±15	5,27299956
Bootes-I-Dwarf	9	196	±9	5,27811466
Ursa-Minor-Dwarf	10	215	±10	5,37063803
Sculptor-Dwarf	11	258	±13	5,55295958
Draco-Dwarf	12	267	±20	5,58724866
Sextans-Dwarf	13	280	±13	5,6347896
Ursa Major I	14	325		5,78382518
Carina-Dwarf	15	329	±16	5,79605775
Hercules-Dwarf	16	430		6,06378521
Fornax-Dwarf	17	450	±26	6,10924758
Canes-Venatici-II	18	490	±49	6,19440539
Leo II	19	669	±39	6,50578406
Canes-Venatici-I	20	718	±82	6,57646957
Leo I	21	815	±82	6,70318811
Phoenix-Dwarf	22	1450	±100	7,27931884
Barnards Galaxy	23	1600		7,37775891
Leo III	24	2250	±325	7,7186855
Tucana-Dwarf	25	2870	±130	7,96206731

The e-function can again be determined from the linearized values. The logarithm distances of galaxies are presented in illustration 8.5.2. The complete e-function for the galaxies is to see in illustration 8.5.4.

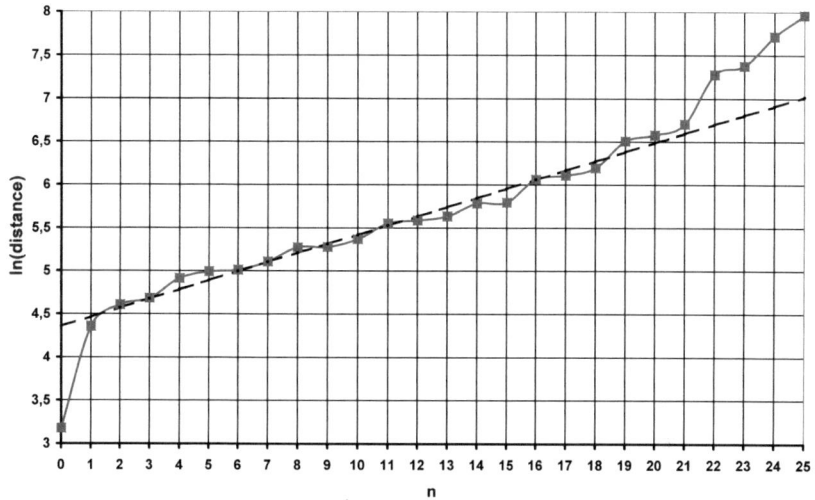

Illustration 8.5.2 – logarithmic galaxy distances

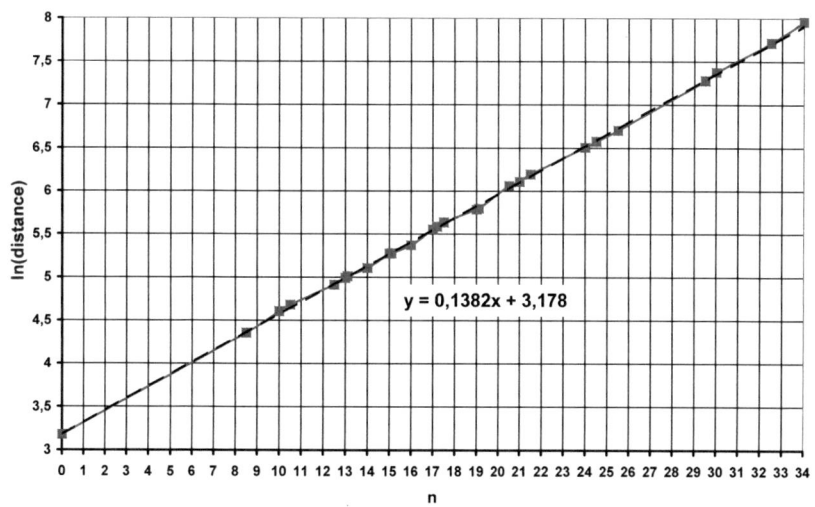

y = 0,1382x + 3,178

Illustration 8.5.3 – linearization

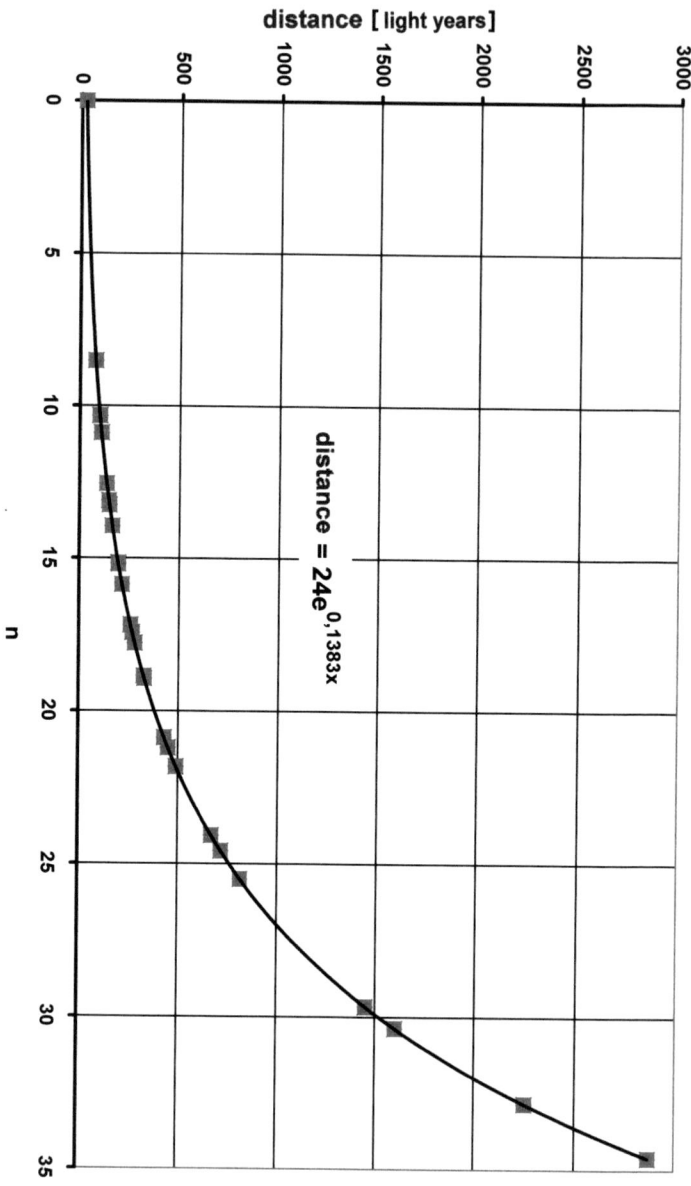

Illustration 8.5.4 – satellite galaxies of the milky way

8.6 – Planetary Nebulae

Planetary nebulae consist of a hull of gas and plasma, which is repelled by an old star at the end of its development. The name is due to historical reasons. The name was therefore that planetary nebulae in the telescope seem mostly round and greenish like gas planets. About 1500 planetary nebulae are known in our galaxy.

Planetary nebulae play an important role in the chemical evolution of a galaxy, because the returning material enriching the interstellar medium with heavy elements such as carbon, nitrogen, oxygen, calcium and other reaction products. A group by their appearance is reminiscent of oscillation structures exist in the planetary nebulae.

Illustr.8.6.1 - Ant-Nebula

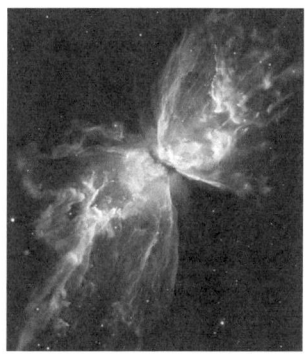
Illustr.8.6.2 - NGC 6302

Illustr.8.6.3 - Boomerang Nebula

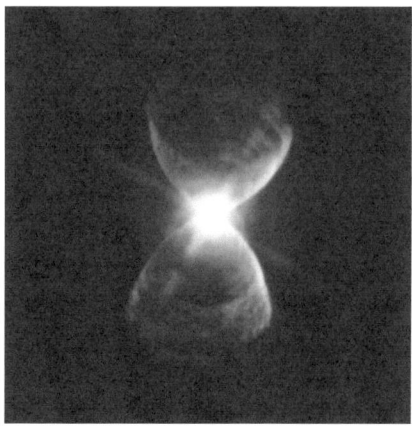
Illustr.8.6.4 - N Hb 12-HS-R658nG656n

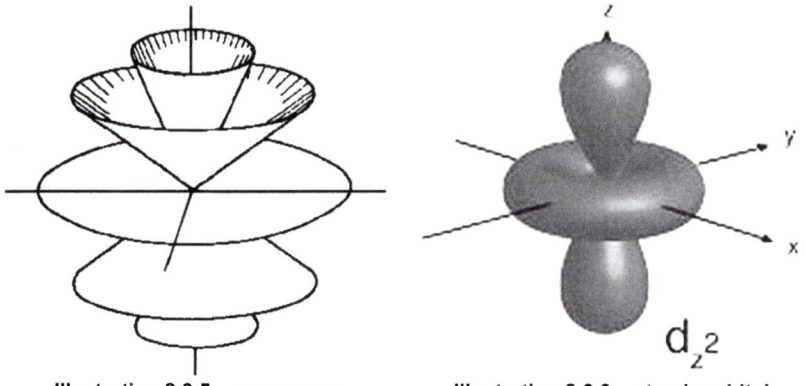

Illustration 8.6.5 – zero areas   Illustration 8.6.6 – atomic orbital

The similarity of the structures allows to conclude that planetary nebulae also oscillation operations take place in. (see also Chapter 2)
That raises the question whether also jet flows or black holes, gamma ray bursts and pulsars can be described as oscillation phenomena.

Another indication of vibration structures in planetary nebulae is a widespread form of planetary nebulae and such that have a concentric structure.

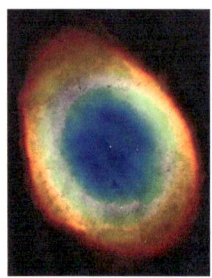  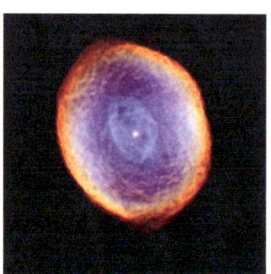

Illustr.8.6.7 - M57   Illustr.8.6.8 - NGC 5979   Illustr.8.6.9-Spirograph

As described in the previous chapter concentricity always is an expression of oscillation phenomena. Therefore it can be assumed in planetary nebulae, that the structure forming and shaping has been designed with radial oscillation.

8.7 – Layers of the earth

Again all the layers of the earth are considered in the light of the described procedures of logarithmic, linearization, and e-function determination.

Nr	Depth	Depth new	Layers of the earth		ln(Depth)
	-6371	1			0
0	-5100	1271	border inner core / outer core		7,147559271
1	-2900	3471	border outer core / mantle		8,152198016
2	-1700	4671	1700 Km discontinuity		8,449128461
3	-1200	5171	1200 Km discontinuity		8,550821372
4	-1000	5371	border mantle / transition zone		8,588769390
5	-920	5451	920 Km discontinuity	900-1080 Km	8,603554357
6	-720	5651	720 Km discontinuity		8,639587800
7	-660	5711	660 Km discontinuity		8,650149419
8	-520	5851	520 Km discontinuity		8,674367866
9	-410	5961	border transition zone / upper mantle		8,692993531
10	-300	6071	X-Diskontinuität	250-350 Km	8,711278615
11	-250	6121	Lehmann discontinuity	190-250 Km	8,719480761
12	-190	6181	Lehmann discontinuity		8,729235350
13	-100	6271	low velocity zone		8,743691111
14	-80	6291	border upper mantle / Lithosphere		8,746875320
15	-60	6311	border Lithosphere / crust		8,750049422
16	0	6371	Earth surface		8,759511722
17	20	6391	Tropopause		8,762646030
18	30	6401	Ozon		8,764209507
19	50	6421	Ozon		8,767329148
20	60	6431	d		8,768885326
21	70	6441	d		8,770439087
22	100	6471	e		8,775085935
23	140	6511	e		8,781248333
24	180	6551	f1		8,787372989
25	200	6571	f1		8,790421307
26	300	6671	f2		8,805525053
27	800	7171	g		8,877800394

In illustration 8.7.1 it is to recognize that most of the data points are nearly linear.

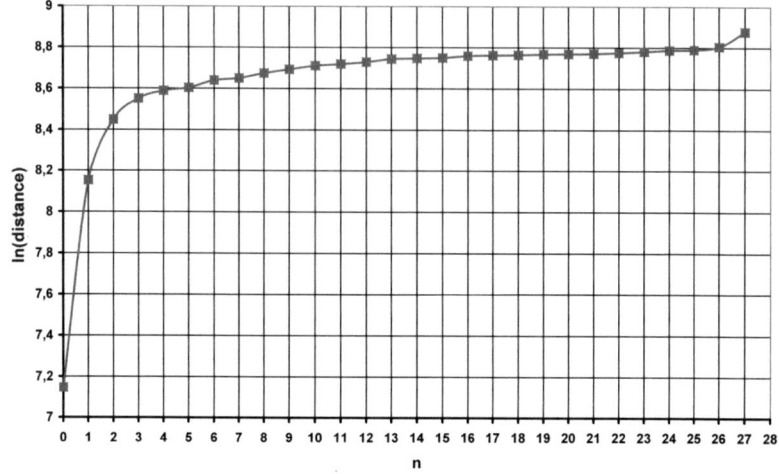

Illustration 8.7.1 – layers of the earth

The slope is so strong for the data points 0, 1, and 2 so they must be moved for a linearization about 1600 counting steps. This is to be seen in illustration 8.7.2.

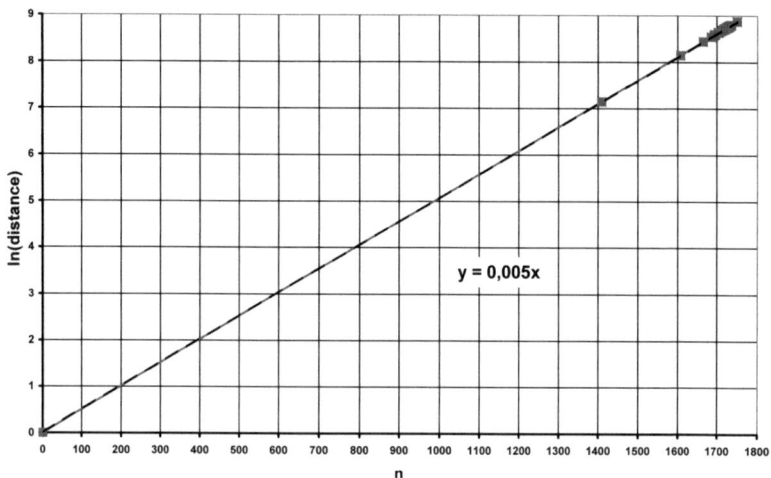

Illustration 8.7.2 – linearization

To note is that the layers in the Earth have a negative sign. Since no logarithms for negative numbers exist, the entire function must so be raised in such way until all values are positive. This is achieved by addition of the earth radius.
As seen the new numbering values can be calculated again, as shown in the following table on page 174. Applies to the derivation of the e-function:

$$\ln(r + R_e) = y = 0{,}005 \cdot x$$

Resolution of the equation yields:

8.7.1 - Equation: $\qquad r = e^{0{,}005 \cdot x} - R_e$

respectively:

8.7.2 - Equation: $\qquad r = e^{\frac{x}{200}} - R_e$

The entire situation is shown in the illustrations 8.7.3 and 8.7.4.

The solution shown in equation 8.7.1 and 8.7.2 is however no solution function of Laplace's equation.
The condition for the two time derivation of the function according to section 2.11.3 is not met here.
Since an e-function derivable for the earth layers and atmospheric layers, which could serve also as the solution of Laplace's equation, the conclusion can be total that both oscillatory systems are indeed compatible to each other, but are both independent oscillation systems as well.

Nr	Depth	Layers of the earth	
0	0		
1408,93	1271	border inner core / outer core	
1606,97	3471	border outer core / mantle	
1665,50	4671	1700 Km discontinuity	
1685,55	5171	1200 Km discontinuity	
1693,03	5371	border mantle / transition zone	
1695,94	5451	920 Km discontinuity	900-1080 Km
1703,04	5651	720 Km discontinuity	
1705,13	5711	660 Km discontinuity	
1709,90	5851	520 Km discontinuity	
1713,57	5961	border transition zone / upper mantle	
1717,18	6071	X- discontinuity	250-350 Km
1718,79	6121	Lehmann discontinuity	190-250 Km
1720,71	6181	Lehmann discontinuity	
1723,56	6271	low velocity zone	
1724,19	6291	border upper mantle / Lithosphere	
1724,82	6311	border Lithosphere / crust	
1726,68	6371	Earth surface	
1727,30	6391	Tropopause	
1727,61	6401	Ozon	
1728,22	6421	Ozon	
1728,53	6431	d	
1728,84	6441	d	
1729,75	6471	e	
1730,97	6511	e	
1732,17	6551	f1	
1732,78	6571	f1	
1735,75	6671	f2	
1750	7171	g	

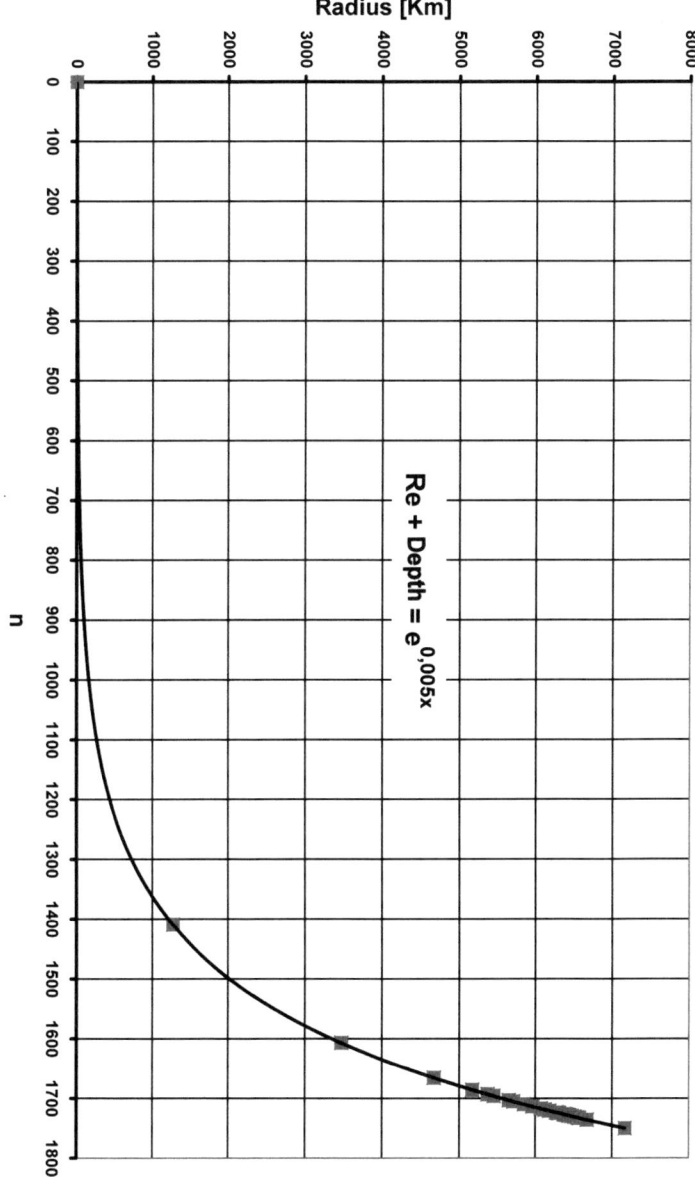

Illustration 8.7.3 – layers of the earth

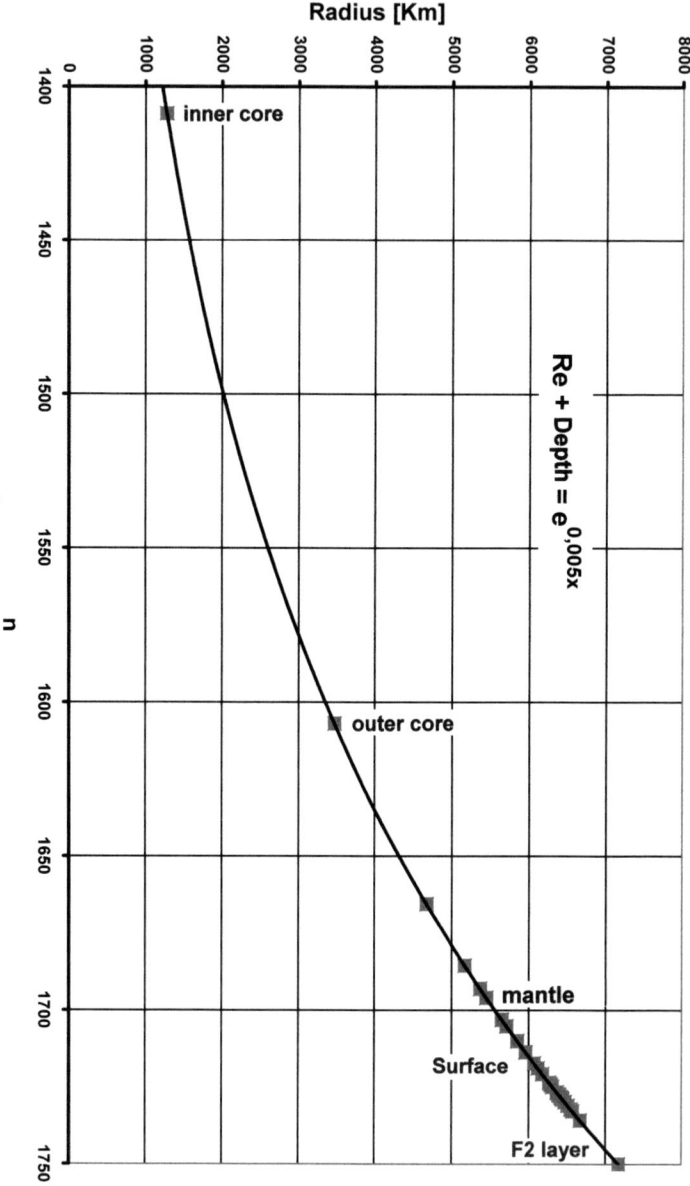

Illustration 8.7.4 – layers of the earth

8.8 – Fruits and flowers

In the biological area, namely the flora, there are examples that have a concentric structure. Below here some fruits and flowers are treated, which meet the requirements of the **concentricity** and multi layers.

8.8.1 – Peach

Fruits are a completely different category (as cosmic objects) of concentric structures. As shown in the adjacent picture the peach has 3 areas - the core, the stone and the skin. Because there are three data points, so an e-function can be generated. In the following table, the data of a peach is to find.

Name	Radius	Nr	ln(Radius)	Nr calculated
Core	0,8	0	-0,22314355	0
Stone	2	1	0,69314718	0,95
Skin	5,5	2	1,70474809	2

By turning the values into logarithmic numbers and transfer into a diagram, like at the moons of Mars results in a simplification. The logarithm values are already so linear that a more linearization is no longer necessary.

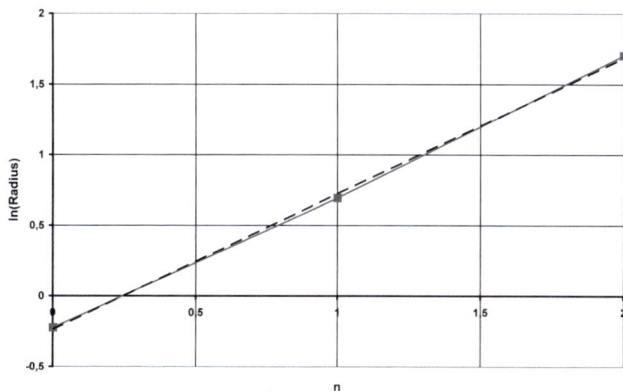

Illustration 8.8.1 – linearization

The entire facts for the peach is to see in illustration 8.8.2.

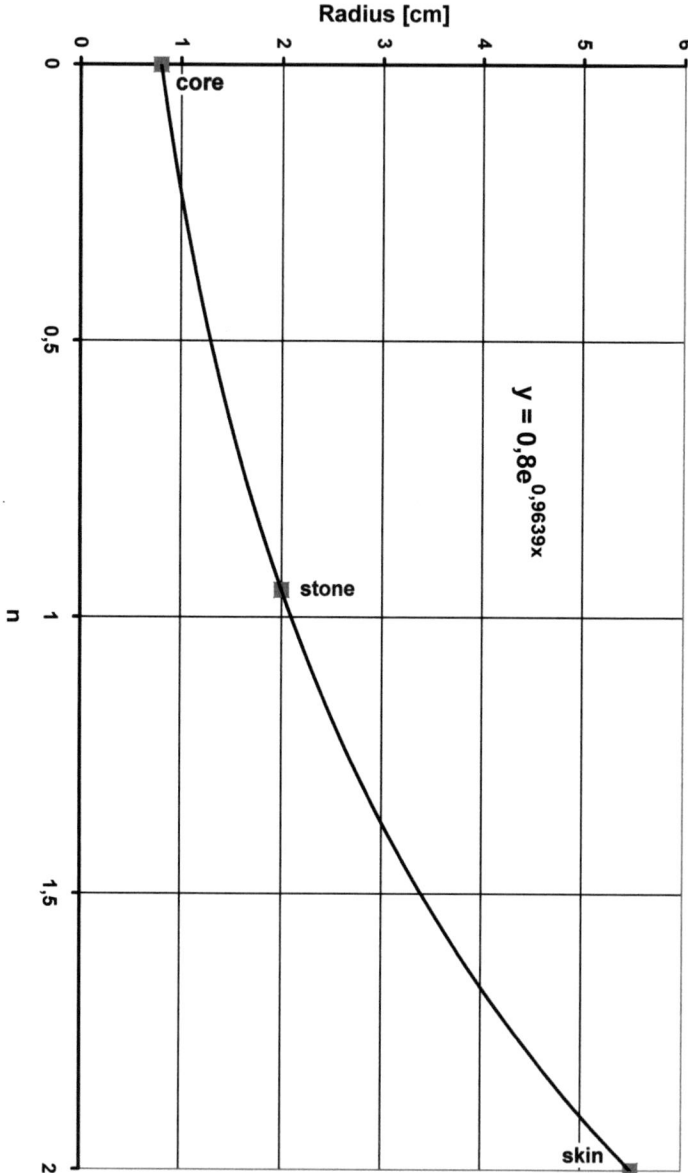

Illustration 8.8.2 – layers of a peach

8.8.2 – Narcissus

As in the adjacent image with the narcissus is to see exist 4 areas.
Because there are four data points, so an e-function can be generated. In the following table are the data of a narcissus.

Name	Radius	Nr	In(Radius)
pistil	0,3	0	-1,2039728
inner area	1	1	0
blossom	2,8	2	1,02961942
leaves	6	3	1,79175947

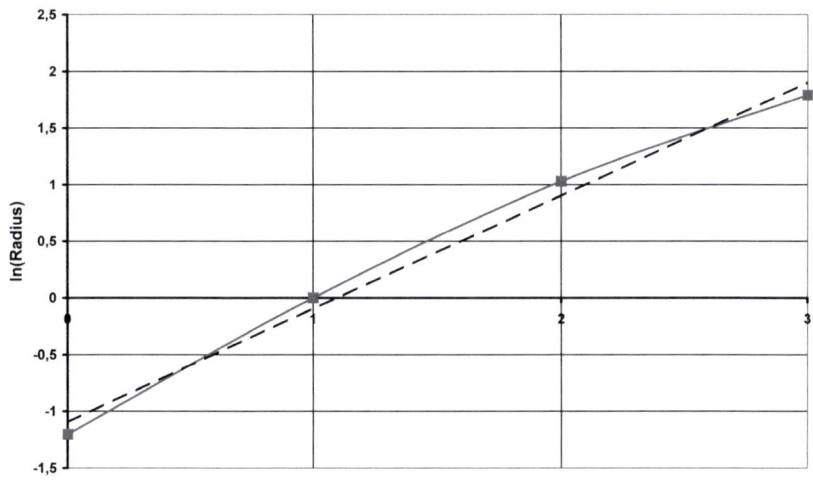

Illustration 8.8.3 – logarithmic radii of a narcissus

Logarithmic and transfer into a diagram, an almost linear function arises similar to as in the peach. An another linearization would be no longer necessary. It is carried out here but for the sake of accuracy.

The linearized function is shown in illustration 8.8.4.

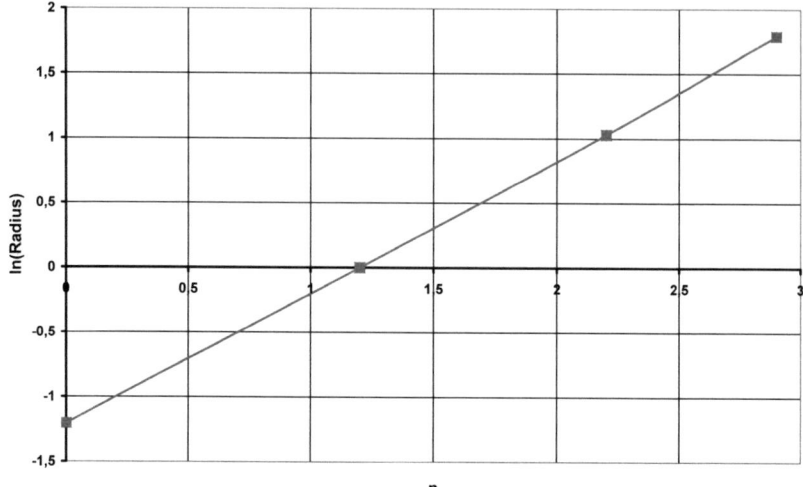

Illustration 8.8.4 – linearization

The e-function can again be determined from the linearized values. As already seen the new numbering values can be calculated again.

Name	Radius	Nr	ln(Radius)	Nr	Nr
				neared	calculated
pistil	0,3	0	-1,2039728	0	0
inner area	1	1	0	1,2	1,21
blossom	2,8	2	1,02961942	2,2	2,24
leaves	6	3	1,79175947	2,9	3

The radii of the narcissus are shown in illustration 8.8.5.

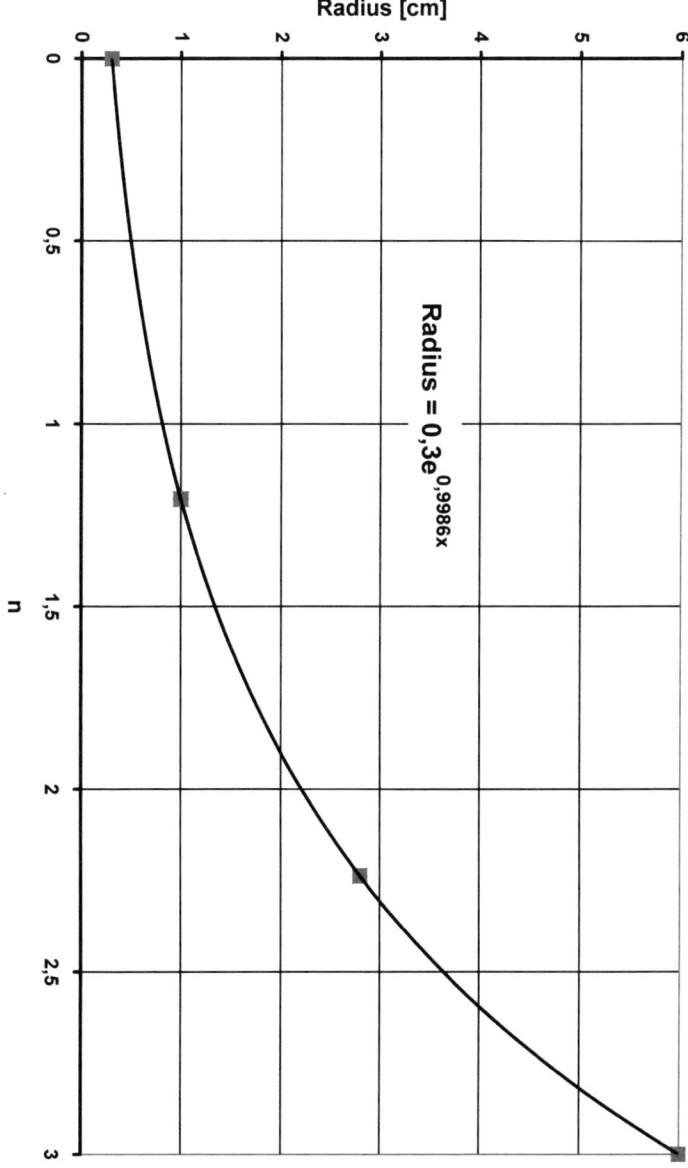

Illustration 8.8.5 – radii of a narcissus

8.9 – Result

The question at the beginning of the chapter was, in how far the other spherical and concentric phenomena of our world can be applied by the derived oscillation model.

e-functions are solutions of the radial component of Laplace's equation. So, you need only to examine whether concentric arrangements in the real world are convertible into an e-function.

With the help of the previously used procedures of logarithmic, linearization, and conversion into an e-function with the equations 7.4.1 to 7.6.2 an algorithm is available that allows the following statement:

8.9.1 - Theorem: Any sequence of numbers ordered ascending or descending (in size) can be converted into an e-function.

The direct consequence of this is:

8.9.2 - Theorem: Each concentric structure can be converted into an e-function.

How to see for the orbits of the planets and the moons of Mars, an amazing linearity already exists in the logarithmic. So an e-function can be seen here. This can be taken as a **strong correlation**.
Because e-functions are solutions for the radial component of Laplace's equation, the following sentence can be formulated:

8.9.3 - Theorem: All concentric structures can be interpreted as solution of the radial component of Laplace's equation.

As in the examples in this chapter, the **linearity of the logarithmic values is a necessary precondition** that the present sequence of values can be converted into an equivalent e-function.
The metric of the numbering, also plays a role. An increment of 1, half, a quarter, etc. a harmonious relationship suggests. A random sequence of numbers would generate even a random metric of numbering.

All the examples here show that concentric arrangements can be interpreted as solution functions of radial spatial oscillation systems.

That is also the message of the global scaling, that the universe has a logarithmic structure, yet again impressively confirmed on the examples.
But the scale factor can be chosen freely. The mathematically simplest solution is to set scale factor equal to 1. Then you will get the absolute harmonic variables in a system.

In the biological area, namely the flora, there are examples that have a concentric structure. The concentricity is met such as fruits like peach, orange, coconut and flowers such as dahlia, the yellow flower (Gerbera) or the narcissus. You also can here create an e-function.

In the face of the entire data, this allows the following conclusion:

8.9.4 - Theorem: **Concentric configurations as solution functions of radial spatial oscillators are an universal design principle for circular or spherical arrangements.**

As solution functions of the Laplace equation, for the radial direction concerning concentric configurations, only the e-function is virtually considered. That turns out an exponential or logarithmic basis for the structuralisation.
Then the consequence of the whole material from chapter 7 and 8 can be formulated in an easy manner so:

8.9.5 - Theorem: **Our universe has a logarithmic respectively exponential structure.**

Epilogue

Here's an epilogue to " the planetary oscillating structures". Is there a reason for such structures ? Can this reason be found in the law of the nature? If yes, this reason must be at least as old as the earth itself, probably much, much older.

Epilogue is Greek and depending on the situation it means: defamation - addendum - final word - extension - summary - addition.

Thereby we already are in the middle of a dilemma - Variations of translations from an ancient Eurasia:
"In the beginning was the word." A well-known quote from the Gospel of John in the Bible, which before it was translated into German, had at least three to four different translations, each with 5-6 meanings. This allows a variety of translations and interpretations, including different meanings in languages such as Aramaic, Greek, Latin and German.

Logos (the word) has in this case the broadest meaning:

A well-known use of the word Logos is found in the Gospel of John in the Bible, which was written in Greek around the year 150 of our time. It strongly resembles the doctrine of Heraclitus, who lived just about 600 years earlier! It begins with these words:

$$Εν ἀρχῇ ἦν ὁ λόγος$$

En archè èn ho logos

In the beginning was the Logos

The usual, but very problematic translation is: "In the beginning was the word." At the latest Goethe took offense for the translation of logos as "the word" (= the outspoken, said). In Faust, he is discussing other, philosophically notable, possibilities:

It's written here: 'In the Beginning was the Word!'
Here I stick already! Who can help me? It's absurd,
Impossible, for me to rate the word so highly
I must try to say it differently
If I'm truly inspired by the Spirit. I find
I've written here: 'In the Beginning was the Mind'.
Let me consider that first sentence,
So my pen won't run on in advance!
Is it Mind that works and creates what's ours?
It should say: 'In the beginning was the Power!'
Yet even while I write the words down,
I'm warned: I'm no closer with these I've found.
The Spirit helps me! I have it now, intact.
And firmly write: 'In the Beginning was the Act!'

Goethe Faust, I, 1232-1235

The original text in the Greek Bible writes "Logos" (not "Word"). The translation of the "this word" means: language, speech, saying, Word, customer, teaching, argument, thought, reason. The school of Stoics, which at the time of the creation of the Bible, was a very widespread school was considering the word "Logos" as a sense of reason or a rationality principle meaning the "original law", as a stationary foundation of the world.
[The only (natural) law / original law]. Therefore, an accurate translation of the word "Logos" does not exist, and "Word" is certainly misleading. God is in this prologue equated with the principle of reason, the principle of controlling the entire universe.

If "the Logos" was in the beginning – meaning the deed or the thought, then perhaps we can apply another translation, namely "the relationship", the reference, the relationship as a "original first principle ", as a " free interpretation " of an **organizing principle**.

In physics as well as in in the whole Universe is **THE RELATIONSHIP**, or rather the organizing principle, the speed of light i.e. the ratio of the path to time.

If -so far unrefuted- Max Planck is right, and every energy only exists in quantum and fixed conditions, then this also applies for all conditions in the microcosm as well as in the macrocosm! If furthermore $E = mc^2$ is valid, and any energy is a product of the mass and the relative speed of light, then it also applies that masses are arranged in a condition of oscillating structures which obeys the state of the quantum.

And – if the word (Logos) was in the beginning, so were in the beginning the relationship, and thereby a given relation to each other, and thus a target for the structure itself.

God = original Principle = Principle of Sense = structure = oscillation
(for the Word was with God .....)

[Because of the choice of the word "God", the author would like to expand the definition of the content to: Original source, first principle, reason-, light- and love principle..]

A quote:
"We" consists to 99.999999% of "Nothing."

How is that?
All bones, proteins, tears, skin, nails, and cells consists of proteins, carbon hydrates and trace elements, which furthermore are composed of mole-

cules, which are composed of atoms and electrons , and these are composed of quarks, bosons, leptons and photons. Loose particles and matter (non-) particles are circling around each other (in fixed space), without touching each other. But also with plenty of space in between them.

So we are made out of "a lot of space in between". And with what is this space filled?
With „nothing"? - As one may ask: " In us is " NOTHING " ? - The answer is: "No way".
The space in between is filled with force fields. And what do force fields do best: they oscillate. These fields provides the right distance and the right relationship to each other: The ratio of protons to neutrons, electrons to core, of attraction and repulsion, from near and far, from large and small as well as from faster and slower.

We will see where that is leading us. As a Summary, addendum as well as an outlook, we can say that we are consisting of oscillating structures in predetermined proportions since the beginning of time.

Best adapted for life in harmony, with and on this planet.

Bibliography

Baumer, Eichmeier   Das natürliche elektromagnetische Impuls-Spektrum
                         der Atmosphäre
                         Archives for metereology, geophysics and
                         Bioclimatology
                         Springer Verlag, Ser A 31, 249-261, 1982

Baumer, Hans   Sferics
                         Die Entdeckung der Wetterstrahlung
                         Rowohlt Verlag, Hamburg, 1987

Berckhemer, Hans   Grundlagen der Geophysik
                         Wissenschaftliche Buchgesellschaft, Darmstadt,
                         1990

Bird, Christopher   Planetary Grid
                         New Age Journal, Mai 1975

Bischof, Marco   Der Kristallplanet
                         Ideengeschichte der globalen Gitternetze
                         Zeitschrift Hagia Chora, Nr.7 (2000/2001) - Nr.19
                         (August 2004)

Brauch Wolfgang, Dreyer Hans-Joachim, Haacke Wolfhart
                         Mathematik für Ingenieure
                         6. revised edition, Teubner Verlag,
                         Stuttgart, 1981

Cousto   Die kosmische Oktave
                         Synthesis Verlag, Essen, 1984

Dobrinski Paul, Krakau Gunter, Vogel Anselm
                         Physik für Ingenieure
                         5. revised edition, Teubner Verlag,
                         Stuttgart, 1980

Gauß C.F., Weber W.   Allgemeine Theorie des Erdmagnetismus
                         Resultate aus den Beobachtungen des magnetischen
                         Vereins im Jahre 1838
                         Eds. C.F. Gauß, W. Weber, 1-57
                         Dieterichsche Buchhandlung, Göttingen, 1839

Gerthsen Christian, Kneser H.O., Vogel Helmut
Physik
13. revised edition, Springer Verlag, 1977

IUGG/IAGA IGRF-1980 / IGRF-2005
International Union of Geodesy and Geophysics
IAGA, the International Association of Geomagnetism
and Aeronomy
http://www.ngdc.noaa.gov/IAGA/

Kittel Charles Einführung in die Festköperphysik
5. revised edition
R. Oldenburg Verlag München Wien 1980

Kittel C., Knight W., Ruderman M., Helmholz C., Moyer B.
Mechanik
Berkeley Physik Kurs 1
Friedr. Vieweg+Sohn Verlag, Braunschweig, 1979

Lundquist C.A., Veis G. Geodetic parameters for a 1966 Smithsonian
Institution Standard Earth
SAO Spec.Rep200, Cambridge/Mass. 1966

McLean S., Macmillan S., Maus S., Lesur V., Thomson A., Dater D.
The US/UK World Magnetic Model for 2005-2010
NOAA Technical Report NESDIS/NGDC-1,
December 2004

Nevanlinna H., Pesonen L.J., Blomster R.
Earth magnetic field charts (IGRF1980)
Geological Survey of Finnland Report,
Q19/22,0/World/1983/1

NGDC WMM-2005
National Geophysical Data Center
http://www.ngdc.noaa.gov/

Piontzik, Klaus Das Magnetfeld der Erde
Zeitschrift für Geobiologie - Wetter, Boden, Mensch
S.35-52, 2-2002

Piontzik, Klaus	Planetare Systeme Zeitschrift für Geobiologie - Wetter, Boden, Mensch 3-2009, 4-2009, 2-2010, 3-2010
Piontzik, Klaus	Gitterstrukturen des Erdmagnetfeldes BOD Verlag, Juli 2007 See also: http://www.pimath.eu/
Piontzik Klaus, Eberrs Gerrit, Jähn Ulrich	Deutsches Patent 102012011759.0 Verfahren zur Messung von magnetischen Wellen
Purcel, Edward M.	Elektrizität und Magnetismus Berkeley Physik Kurs 2 Friedr. Vieweg+Sohn Verlag, Braunschweig, 1979
Schumann, W.O.	Über die strahlungslosen Eigenschwingungen einer leitenden Kugel, die von einer Luftschicht und einer Ionosphärenhülle umgeben ist Zeitschrift Naturforschung 7a, 149-154, 1954
Schumann, W.O.	Über elektrische Eigenschwindungen der Hohlraumes Erd-Luft-Ionosphäre, erregt durch Blitzentladungen 1957, Zeitschrift Angew. J Physik 9:373–378
Stieglitz R., Müller U.	Kann man das Magnetfeld im Labor simulieren? Forschungszentrum Karlsruhe Wissenschaftliche Berichte, FZKA 6223, 1999
Torge, Wolfgang	Geodäsie 1975, Walter de Gruyter, Berlin, New York Sammlung Göschen 2163

Picture credits

Illustr.1.0	NASA
Illustr.2.2.5	Physik Gerthsen, Kneser, Vogel, 13. Edition, Page 120
Illustr. 2.7.3, 8.6.6	http://www.ck12.org/ High School Chemistry.pdf, Page 271
Illustr.4.0.2	Physik Gerthsen, Kneser, Vogel, 13. Edition, Page 256
Illustr.4.1.1	Grundlagen der Geophysik Berckhemer Hans, Page 138
Illustr.4.2.1	The US/UK World Magnetic Model for 2005-2010 Page 15
Illustr.4.3.1, 4.3.2	The US/UK World Magnetic Model for 2005-2010 NOAA Technical Report NESDIS/NGDC-1, Dezember 2004 Page 49
Illustr.4.4.1	World Data Center for Geomagnetism, Kyoto http://swdcwww.kugi.kyoto-u.ac.jp/
Illustr.8.6.1-8.6.4	NASA, ESA & the Hubble Heritage Team (STScI/AURA) The Hubble Heritage Team (AURA/STScI/NASA)
Illustr.8.6.7-8.6.9	NASA, ESA & the Hubble Heritage Team (STScI/AURA) The Hubble Heritage Team (AURA/STScI/NASA)

All other Illustrtrations come from the archives of the authors.